Also by Orest Bedrij

Yes It's Love: Your Life Can Be A Miracle, ed. (1974, 2009)

One (1977, 1978)

You (1988)

Chto Ty Ye (Who Are You) (1998)

La Preuve Scientifique de L'Existence de Dieu (Science Proves Existence of God) (2000)

Seeing God Face to Face (2004)

Celebrate Your Divinity: The Nature of God and the Theory of Everything (2005, 2007)

'1': The Foundation and Mathematization of Physics (2008)

Living Your Divine Life: Experience God's Glory, Absolute Happiness, and Great Prosperity (2009)

Exodus III (2011, 2013)

'1': The Foundation, Prediction, Verification, and Mathematization of Pure Being (2014)

The Greatest Achievement: Miracle after Miracle the Easy Way (2014)

'1': The Encyclopedia of Physical Laws (Vol. 1, 2, 3, 2016)

Your Miracle after Miracle Life
Celebrate Your Essence, Celebrate Your Eternity

My Absolute Splendor—the Light of compassion.
You,
Manifesting in countless ways and forms.
You,
In all Your beauty, wisdom, peace, and harmony.

Your Miracle after Miracle Life
Celebrate Your Essence, Celebrate Your Eternity

Orest Bedrij

Print information available on the last page.

Rev. date: 10/19/2020

To order additional copies of this book, contact:
Xlibris
844-714-8691
www.Xlibris.com
Orders@Xlibris.com
795416

Jacob's Ladder Leading to the Eternal
The House of the Beloved
and the Gate of Heaven

From unawareness, we advance to belief.
Through belief, we realize knowledge and understanding.
With knowledge and understanding, we achieve wisdom.
Wisdom in action is freedom, beauty, love, and compassion.
Freedom, beauty, love, and compassion in action are God visible.

CONTENTS

Hosanna! Thank You, Father

O Father, my Sweetheart, my Essence and my Love,
Oh, Infinite Wisdom and Light,
Immeasurable love to you
For letting me experience you
In freedom, peace, and kindness;
Excellence, choice, and humility;
Teamwork, gentleness, and responsibility;
Simplicity, longer and abundant life;
My precious parents;
My dearly beloved wife and children;
My treasured coworkers and neighbors.

O Great Joy, Goodness, and Happiness
Oh, Holy Compassion and Care—
When I contemplate your nature,
Words are incapable to describe your
Vastness, beauty, perfection, and eternity;
Only the music of purity, service-excellence, and silence . . .

O Bliss of bliss and Door of doors,
Oh, clear Light of light in your Being,
Please allow me to love you and glorify you,
With all my heart, all my soul, and all my being.

Glory to you Grace, Glory to you Father,
And on earth peace and goodwill toward all.
Thank you, Father, as Jesus the Christ,
For revealing our divinity and the way to you back in eternity.
Thank you, Father, as Mother of Perpetual Help Mary,
For the Transfiguration and Resurrection of humanity.
Thank you, Father—as Chrystyna Stecyk, Roksana Bedrij Arpa, and
Christiaan Bedrij Arpa, David Beatty, Narlyn Dura, Andy Ferrer, and
Olivia Henderson—For your editing help.
I love you.

Orest

PROLOGUE

Your Peace, Joy, and Your Miracle after Miracle Life

My dearest reader, I love you most dearly!

May grace, harmony, and tranquility be yours in full measure by your experience of God's divine peace, which passes all understanding; the sanctity of your Being, which lives in us; your miracle after miracle life; and your precious Holy Family!

This work offers a new dimension with infinite mapping, through which to experience your miracle after miracle life, liberty, and the pursuit of happiness. It employs the ultimate principle—the *final law* ('**1**') of the universe, which is the *nature of God* and is higher than and greater than the fundamental laws of nature (i.e., of space and time and the theory of everything).

The final law is the "I AM" ('**1**', in all places) and which is known as the "I AM WHO I AM" of Exodus 3.14 that God recognized (acknowledged, made known) *Himself* to Moses and humanity. To be exact: "Thou art That," which is explained more completely and *which is you*.

The ultimate principle (the Eternal Essence, the Christ Nature), which is *above* the fundamental laws of nature, does not require clarification or elucidation in terms of deeper principles as well as the unsolved problems in science, belief, values, or government. This work is for readers with no prior knowledge of philosophy, spiritual conviction, higher mathematics, or physics. It is beyond the standard model,

M-theory, supersymmetry, quantum gravity, and general relativity. At this juncture, the foundations of physical science come to a revered pause.

We have incorporated the knowledge of modern physics, science, and mathematics into the sacred and secret understanding of the nature of God (the Being of beings, His eternity and immutability) in Buddhism, Christianity, Hinduism, Islam, Judaism, etc., with personal direct experience and verification. We hold within us ('=') the miracles we search for outside us.

To live a holy life of oneness in liberty and the persuit of happiness is sacred, blessed, and righteous. Here and now, you will be able to experience the deep *inmost* Holy Spirit with innermost transfiguration, resurrection, and ascension: the highlight is on the "I AM" ('**1**'), Lord of Light and "thou art That" technology that you can actually see, understand and experience *beyond the known* of deepest purity, silence, stillness, peace, with rest (see appendix 1). To live in the God-realm ('**1**') beyond the known and human thinking ('='), while thought is restricted to realm of the known. Crowned with glory and honor, and your consecrated Holy of Holies (the deepest, most hallowed and most revered) world will be for you in the making.

"God is love, and he who abides [lives] in love abides [lives] in God, and *God in him*" (significance added), wrote Saint John (1 Jn 4:16). Now you are the Absolute Splendor on earth in rapture motion. "He became man that we might be made God,"[1] said Saint Athanasius the Great (296–373), Pillar of the Church and champion of Christ's Divinity. "God became man, and as a result, the whole human family has been raised up and ennobled,"[2] affirms Dominican priest, preacher, and theologian Meister Eckhart von Hochheim (1260–1328).

Furthermore, God ("I AM" [Ex 3:14][3]), the Supreme Glory, has not only become human family ("I AM THAT I AM"); but more so, he/she is full of human nature, beauty, harmony, wisdom, compassion with venerated and creative miracle-after-miracle knowledge, life, and power from eternity on earth. God is also observable life and the uncreated Light—a vast ocean with infinite potential for endless discoveries of beauty, love, and joyful adventure.

Because the secret doctrines of Christianity (concealed wisdom of Christianity is found all through the early church), the unqualified (uninitiated)—with shortcomings and imperfections—were not given "solid food" (more advanced Christian knowledge, wisdom, and experience on which they could literally choke) about humanity's I AM origin. It's a beautiful face of love, fire and wonder, but mother's or infant's "milk" (Christian metaphor for the *insufficiently prepared* new believers) (1 Cor 3:1–3). They were not permitted to know, see, study, and meet (behold) on sacred ground. *The forbidden knowledge,* the most safeguarded secret of the ancients: *that the Most High Himself takes on human form and lives in the world as us,* "The Word became flesh [a human being] and made his dwelling [home and living] among us" (Jn 1:14). And the secret is this: "the mystery of *Christ in you,* your hope of glory, which means that *you will share the glory of God*" [my accent] (Col 1:27).

For example, Saint Basil the Great (329–379), one of the most distinguished doctors of the church, also known as the revealer of heavenly mysteries, the bishop of Caesarea Mazaca in Cappadocia, Asia Minor (370–379), instructed his various Eastern Churches that "the uninitiated [unqualified] are not permitted to behold these things. Their meaning is not to be divulged by writing it down."[4] The "solid food," for Saint Basil the Great, was this:

> God [I AM, That Which Is] *is on Earth,* He is among men ["I AM THAT I AM"], not in the fire nor amid the sound of trumpets; not in the smoking mountain, or in the darkness, or in the terrible and roaring tempest giving the Law, but *manifested in the flesh* [my emphasis], the gentle and good One dwells with those He condescends to make His equals![5]

The Christian Church has four canonical Gospels, usually credited to two apostles, Matthew and John; and to two disciples of apostles, Mark and Luke. These Gospels were selected out of a larger number of books that were in readership. There were many heated debates and discussions as to which books are the only true and authentic ones. The heated debates and discussions lasted over three hundred years. These four canonical Gospels, lost scriptures, suppressed, forgotten, or destroyed scriptures (list of some books that did not make it into the New Testament[6]) *and* the holy scriptures of other religions and traditions are *only* road maps (preparation for a declaration of independence). Road maps show how to *connect* and *synchronize* with your Source and your Being, our Father. Think about that.

Once you bond with God, once you achieve *union without ceasing* (see appendix 1), you let God be God in you. You formally proclaim youself as a free and independent life. Like Saint Paul in his Epistle to the Galatians verbalized, "The life I live now, is not my own; Christ is living in me" (Gal 2:20). Or as Willis Harman in this book states, the business of the stomach is not the growth of the stomach, but the health of the body.

Christianity grew out of Judaism. Jesus, the apostles and the evangelists were Jewish. The Israelites wrote their holy text in Hebrew and Aramaic, which are sister languages. Jesus's references and quotes from the Old Testament see eye to eye with the Peshitta text (in Aramaic the original and pure New Testament) but do not agree with the Greek text. For illustration, in John 12:40, the Peshitta Old Testament and New Testament agree.

The New Testament were initially written in Aramaic. Aramaic Gospels were then translated into Greek. From Greek, the Gospels were translated into Vetus Latina ("Old Latin" in Latin), then into Vulgate, the Latin translation produced by Jerome 382-405, and afterward in 1535 first complete English Bible.

Aramaic primacy contrasts with existing consensus that initial language of the New Testament was Greek. Many English intermediary translations don't correspond to the original *deep* and *subtle* Aramaic as well as Cosmic Consciousness (Christ Nature) meaning of Jesus. As an example, we will address for you Jesus's at the ninth hour statement on the cross: "Eloi, Eloi lama sabachthani" (Mk 15:34) in Aramaic and in English Christ Nature inescrible importance and significance translation. Jesus's (Aramaic) to English New Testament translation emerges to be the best without going to Greek and Latin to English road.

Note: typical consciousness deals with simple knowledge of everyday experience—that is, knowledge and consciousness, where most people live and have their consciousness home. Cosmic or Christ Consciousness is a higher, multidimensional, form of consciousness and knowledge (namely: seeing a 3-D color motion picture film versus a single black and white picture), which we will present shortly. Cosmic

Consciousness, or Christ Nature, deals with multilayered, bottomless hologram knowledge, which one can simultaneously experience. To be exact: what *is*, what *was* in the past, and what *will be* in the future, as in the transfiguration (Mt 17:1–8; Mk 9:2–8; Lk 9:28–36; 2 Pt 1:16–18) foreknowledge of God's purpose by dying in Jerusalem and the meaning of the Resurrection from the dead (Mk (9:9-11).

"The Lord God says, 'I am the Alpha and the Omega, the One who is, who was, and who is to come, the Almighty!'" (Rv 1:8). Who is, who was, or who will be in the future exists in mathematical form, *simultaneously*, like a movie—from an initial state of a zero point, the singularity ('**1**'), of the big bang (how the universe began) through its subsequent large-scale advance and development (cosmic or cosmological inflation, and Universal evolution from subatomic particles to human society) to following contraction of the big crunch (how the universe will end) zero-point singularity.

Jesus can view, examine, and experience what *was* in the past or *is* to come (foreknowledge) in the future:

a) Peter's denial (Jn 18:15–27): before the rooster crows that morning Peter would deny Jesus three times: "And he began to curse and to swear, saying, I do not know this man of whom you speak" (Mk 15:71, Lamsa).

b) The Samaritan woman at the well (Jn 4:4–42): "You are right in saying you have no husband!" Jesus exclaimed. "The fact is, you have had five, and the man you are living with now is not your husband. What you said is true" (Jn 4:17–18).

c) Judas's betrayal (Jn 6:64, 13:28–28).

d) Before the arrest, "the end has arrived and the hour has come; and behold, the Son of man will be delivered into the hands of sinners. Arise, let us go; behold, he who is to deliver me is near. While he was speaking, Judah of Iscariot, one of the twelve, and many other people, came with swords and staves, from the high priests and scribes and elders," (Mk 14:41-43).

e) "When I am risen (in Jerusalem), I will be in Galilee (the area that is north of the Mount Carmel-Mount Gilboa ridge) before you," (Mk 14:28, Lamsa).

f) Jesus observed and shared his suffering, death, resurrection, ascension, destruction of the Second Temple in Jerusalem 70 CE (the Roman army captured the city of Jerusalem and destroyed both the city and the Temple), persecution and annihilation of the apostles and disciples, that nation will rise against nation, there will be plagues, terrors, famines, etc. (Mt 24:1-51, Mk 13:1-37, Lk 21:3-36):

> From then on Jesus started to indicate to his disciples that he must go to Jerusalem and suffer greatly there at the hands of the elders, the chief priests, and the scribes [of the Law], and be put to death, and raised up [to life] on the third day. (Mt 16:21)

Arthur Eddington ("We have succeeded in reconstructing the creature that made the footprint. It is our own"), Max Planck ("I regard consciousness as fundamental"), John Wheeler ("The universe does not exist 'out there' independent of us"), Niels Bohr ("We are ourselves both actors and spectators"), David Bohm, Erwin Schrödinger, and Abraham Joshua Heschel ("Ultimate embarrassment") present now different aspects of your miracle after miracle life.

Similar to a 3-D motion-picture movie, it can be accessed via *infused contemplation*, which is discussed in this work, and how you can apply it in your life, work, creativity, stock or currency selection, forward thinking, etc. I have utilized infused contemplation in the development of supercomputers, high precision unification of physics, and generation of the fundamental physical constants of very high precision (see appendix 2).

Like Jesus Christ, the Buddha, Eddington, Planck, Wheeler, Bohr, and others (Wolfgang Mozart, Ludwig Beethoven, Johannn Sebastian Bach, Nikola Tesla, etc.), you can also observe, study, research, and understand what was in the past or is to come in the future. That is, *solve* global warming, environmental pollution, famine, drought and pandemics; eliminate economic, racial, and social inequality; improve agriculture, aquaculture, ecology and sustainability research; and perfect technology and other fields like gravity propulsion (search on the Internet: Bedrij and gravity physics), through infused contemplation.

Reminder: Christianity is a *revelation of mysteries* in religion (Eph 1:9–10, 3:8–11, Col 1:25–27, Rv 1:20) of how God's secret design and hologram arrangement is put into realization. In the Greek writing, *mystery* ("musterion") represents a mystery, secret doctrine, and hidden knowledge revealed only to the initiated.

The term *mysticism* has Ancient Greek origins referring to the understanding (decoding) of hidden or ultimate truth. In contemporary era, mysticism has attained a narrow meaning as the aim at the *union* with God or the Absolute (see appendix 1). Mysticism can be found in each and every religious tradition, including aboriginal traditions and shamanism. A mystic is one who seeks to decode and understand the mystery of the Absolute by *synchronization*, self-surrender, and contemplation by way of the Absolute Light of Certainty.

Muhyiddin ibn Arabi (1165–1240) was an Arab Andalusian Muslin mystic, poet, scholar, and a holy man whose *unity of being* (the One Mind, the Buddha Mind) insights and writings has become very significant and noteworthy outside the Muslim populace. Like Jesus and the Buddha, he realized that *only* God exists. God is the One who is underlying the many and is also the many. Ibn Arabi was a practitioner of Sufism and lived in Mecca for almost three years, where he was known as the *Greatest Master* and revivifier of religion. "There is no other existence than He," ibn Arabi writes:

> When the mystery [here in Christianity, Judaism, Hinduism, Buddhism, etc.]—of realizing that the mystic is one with the divine—is *revealed* to you, you will understand that you are no other than God [the kingdom of God is within you; "I AM That I AM"; I am that; That which permeates all]; and that you have continued and will continue . . . without when and without times. Then you will see all your actions to be his actions and all your attributes to be his attributes and your essence to be his essence, though you do not thereby become he or he you, in either the greatest or the least degree. 'Everything is perishing save his Face,' that is, there is nothing except his Face, 'then, whithersoever you turn, there is the Face of God.'[7]

The Upanishads are the oldest Sanskrit texts of spiritual knowledge of Hinduism, the Vedas, and the most important literature in the history of Indian religions and culture. Like ibn Arabi in the Muslim humanity ("There is no other existence than He"), Upanishads state, "All the world is Brahman" and "This [whatever exists], too, is Brahman."[8] Brahman, Sanskrit, is the creator god (self-born, the creative aspect, god of Vedas) in Hinduism.

Similarly, Taoism, the philosophical tradition of Chinese origin that emphasizes living in harmony with the Tao (the '1'), insists that "there is nothing outside the Tao; you cannot deviate from it."[9] Specifically, everything is present within yourself. Search for nothing beyond of yourself.

In the Gospel of Thomas, Jesus said, "I am the light that is above them all. I am the all; the all came forth from me, and all attained to me. Cleave a [piece of] wood; I am there. Raise up a stone, and you will find me there."[10] In Luke, Jesus stated, "To you the mysteries of the reign of God have been confided, but to the rest in parables that, 'Seeing they may not perceive, and hearing they may not understand'" (Lk 8:10). Specifically, the information of the secrets (mystery) of the kingdom of God has been given to you; but to the rest, it comes by means of parables and allegories so that they may look but not see and listen but not appreciate. The reason for the secrets (mystery) and parables, we see this in Matthew: "Do not give what is holy to dogs, or toss your pearls before swine [the ignorant and unenlightened]. They will trample them under foot, at best, and perhaps even tear you to shreds" (Mt 7:6).

Zoroaster (628–551 BC), Mahavira (599/540–527/468 BC), Pythagoras (570–495 BC), Buddha (480–400 BC), Lao Tzu (sixth century—fourth century BC), and the like had comparable venerated guiding principle. Similarly, in today's world—for sensitive information in government, corporate, and private dealings—we utilize security clearances and need-to-know authorization.

No religion, no country, no man, and no woman is "an island unto itself." We are one holy family. We become conscious that all we encounter, pleasant or unpleasant, is the Beloved coming into touch with the Beloved within us. We recognize that human life is sacred. It must be venerated and celebrated. No one has the right to kill either oneself or others. We realize that we must correct ourselves before life does it for us.

This work can advance your being to know and experience your true nature and facilitate miracle-after-miracle thinking and life. It will be a breakthrough in your consciousness and a second birth (Jn 3:3–7, 1 Pt 1:23). The greatly revered and honored utter transmutation of the self in the Beloved: the state of God-consciousness, characterized by unlimited understanding, purity of intent and love with super normal vitality, formation of a higher state of cosmic awareness, great divine joy, profound inner peace, synchronization, and insight. You no longer will have to make *inquiries* to learn when the time in power of God will come. Neither is it a question of reporting that it is here or there. "The reign of God is already in your midst" (Lk 17:21), as the greatest writer of the age (*War and Peace, Anna Karenina, Resurrection,* etc.) Leo Tolstoy (1828–1910) did in his book, *The Kingdom of God Is Within You.* You will experience the Christ Nature within yourself and live in the midst of miracles-after-miracles directly and absolutely.

I speak from firsthand knowledge, direct personal experience, verification, and corroboration from my heart to your heart. The actual essence and heartfelt significance are about your fundamental nature, your highest state of consciousness, and about your realization and experiencing miracle after miracle without doing—that is, experiencing miracle after miracle the easy way: by the providence of God—with pure science basic for life, with 100 percent science necessary for understanding God in yourself.

We are sons and daughters of God (Jn 1:13). *We have divine blood within us.* We are on a great intellectual adventure, the journey to our transformative essence, humility, and love. And thus, we should wake up from the dead—"wake up and get up from our mortal slumber by an adventurous exploration of the mysteries that lie behind mysteries and of the core behind the core of things"[11] Meister Eckhart von Hochheim shares in Sermon Six. In its full significance, it stands for the advancement of yourself in which all of your thoughts, desires, actions, events, dealings, and measures are in hallowed alignment—*synchronized* with the ultimate inviolable principle (Being of beings) that is above the laws of nature: the unbreakable mind of God. In essence, *he who wants to know God has to know himself* (appendix 1).

Life, liberty, and the pursuit of happiness is a learning process from our first to our last breath. Also, knowing that *all life, liberty and the pursuit of happiness is sacred* and that the kingdom of God is within you, you let the mind of God—the sacred mind of simplicity, purity of heart, compassion, holy peace, and selfless service, which is beyond all phenomena—govern its kingdom. *God does not need advisers.* Let God be God. In other words, you let that infinite reality express itself. Moreover, with the founding of the self as a might and wisdom and contribution in eternal life you support the *interests, well-being,* and *supremacy* of the Absolute.

Be in love. When you surrender yourself and acknowledge yourself as simply an instrument of the miracle providence and might, that hallowed Being greater than you—that sacred and holy essence, humility, and simplicity—will take over your dealings and affairs along with the fruits of actions. You are no longer subdued and inhibited by them, and the work goes unimpeded. You experience miracle after miracle the easy way. So from your heart, celebrate your eternity.

Jesus's and Paul the Apostle's lives reflect this venerated type of existence: The I AM, "I am who I am," and "Thy will be done" method. Abraham Lincoln (1809–1865), the sixteenth president of the United States, with 15,000 books written about him, called this method or this holy approach the providence of God. Please note his insights and wisdom below. Result: You "become sharers of the divine nature" (2 Pt 1:4). No uncertainties for you. No fears for you. No doubts for you. Heaven on earth is in your being.

Saint Paul (AD 5–67) was a contemporary of Jesus (4 BC–AD 30/33). To begin with, Saul (now called Paul) of Tarsus was a persecutor to the death (Saint Stephen [AD 5–AD 34], first martyr of Christianity [Acts 7:58–60]) and violent threats of murder against the disciples of Jesus (Phi 3:6; Acts 9:1) who, after conversion and enlightenment (Acts 9:18), became one of them (Acts 9:1–9). Saint Paul understood, in a flash, that Jesus brought an update and modernized, simplified, and restructured the Law of Moses or Torah of Moses[12] to which he, as a Pharisee, was devoted. He saw that the Messiah (savior or liberator of his people) for which Israel was waiting has come. Furthermore, the *Messiah is in every human being*.[13] Also, Saint Paul has become conscious that it is no longer "I that live": "The life I live now is not my own; Christ is living in me" (Gal 2:20). He integrated and incorporated that Absolute Essence, that I AM, within his being.

Savior Jesus Christ, to Him is the glory now and forever, did the same with his life when, on Mount of Olives, he stated, "Father, if it is your will, take this cup from me; *yet not my will, but yours be done* [emphasis mine]" (Lk 22:42). The outcome is the transfiguration of Jesus Christ, resurrection from the dead, ascension with his material body, *raising and ennobling the whole human family*, and facilitating the establishment of Christianity—freeing the lost, which we will discuss shortly.

Nearly two thousand years later—September 22, 1862—almost like a national saint, President Abraham Lincoln constantly turned to *yet not my will, but yours be done*. He turned to the providence of God ("That power can use me or not use me in any manner, and at any time, as in His wisdom and might may be pleasing to Him") as a basis of "the will of providence in this matter (issued the preliminary Emancipation Proclamation in the midst of Civil War)" to elucidate and to support and to justify why the war started and refused to go away:

> I hold myself in my present position and with the authority vested in me as an instrument of Providence. I have my own views and purposes, I have my convictions of duty, and my notions of what is right to be done. But I am conscious every moment that all I am and all I have is subject to the control of a Higher Power, and that Power can use me or not use me in any manner, and at any time, as in His wisdom and might may be pleasing to Him.[14]

To a gathering of ministers, President Lincoln acknowledged his confidence in the prevailing providence and how it imparted support to his executive affairs:

> If it were not for my firm belief in an overruling Providence, it would be difficult for me, in the midst of such complications of affairs, to keep my reason on its seat. But I am confident that the Almighty has His plans, and will work them out; and, whether we see it or not, they will be the best for us.[15]

This is the significance of Lincoln's self-taught approach: the United States of America and a model of freedom, liberty, prosperity, happiness, and wonder for all people on earth. The doors are wide open. Your heart knows the way. Run, run, and reveal the divine nature and God's heavenly reign in your eternity. As the American music writer Irving Berlin declared, "God bless America, land that I love / Stand beside her and guide her / Through the night with the light from above."[16]

Where Are We Now?
God to Man and Man to God

You Have Slept for Millions
and Millions of Years

Friend, wake up! Why do you go on sleeping?
  The night is over—do you want to lose the day the same way?
Other women who managed to get up early have
  already found an elephant or a jewel . . .
So much was lost already while you slept . . .
And that was so unnecessary!

My inside, listen to me, the greatest spirit,
  the Teacher, is near,
wake up, wake up!

Run to his feet—
he is standing close to your head right now.
You have slept for millions and millions of years.
Why not wake up this morning? . . .

Oh friend, I love you, think this over
carefully! If you are in love,
then why are you asleep?[1]

  Kabir

The Story of the Hidden Treasure
Awakening to a New Level of Reality

One of the most profound human expeditions in life is the search for God. Some people advance further. Their lifelong quest is to see God face-to-face or experience the living God directly, exactly, and openly. In the *Mathnavi*, Rumi recommends that "whoever enters the Way [to God] without a guide will take a hundred years to travel a two-day journey."[1]

"What I want to achieve—what I have been striving and pining to achieve these thirty years—is self-realization, to see God face to face," Mahatma Gandhi puts in writing. "All that I do by way of speaking and writing, and all my ventures in the political field, are directed to the same end."[2] This is the story of the hidden treasure within you: to enter into a spiritual union with the divine absolutely, completely, and truthfully (appendix 1).

Over forty-five years ago, I had the honor to address a large audience of nuns, monks, priests, and students on my new book, *Yes, It's Love: Your Life Can Be a Miracle*. My opening question to the audience, by show of hands, was to ask, "Who believes in God?" As far as I could tell, every person raised their hand, including the trustees on the podium of the auditorium. I also wanted to know if there was anybody in the audience who did not believe in God. There were no hands in the air. At that time, I made a statement: "I do not believe in God." There was unrest and worry on the podium and in the lecture hall! I gained their attention and continued with this story of the hidden treasure.

There was a poverty-stricken destitute family of a husband, wife, mother-in-law, and eight children who had their home in a small two-room log cabin in Kentucky. One day, a petroleum exploration engineer from a major oil company came by searching for hydrocarbons, which could be either crude oil or natural gas. Because the family had a patch of land, the company asked for permission to evaluate their land's petroleum geology, formation assessment, drilling economics, well engineering, reservoir simulation, and oil and gas production potential. After an in-depth analysis and a screenshot of a structure map generated by contour-mapping software for an 11,500-foot-deep gas-and-oil lake, the oil company and the penniless family, based on the geophysics study, *believed* there was oil and gas on the property. They signed leases and started drilling. When the drilling commenced, the oil company expected natural gas and a small amount of oil. Instead, there was a large blowout geyser, which overloaded the storage tanks. Now everybody stopped believing that there was oil on the property.

Through Belief We Realize Knowledge Immeasurable Value of the Kingdom of Heaven

Here I informed my listeners that *I know God*. Also, by show of hands, I wanted to see who in the audience understands the deepest meaning of today's scriptures, who knows God. There were no hands. The hidden nature of the spiritual treasure and no show of hands indicated to me that the Absolute Splendor, the kingdom of heaven, is not yet revealed to everyone. The unknown God who people worship has not revealed Himself in us (Acts 17:23),

> For the God who made the world and all that is in it, the Lord of heaven and earth does not dwell in sanctuaries made by human hands; nor does he receive man's service as if he were in need of it. Rather, it is he who gives us all life and breath and everything else. (Acts 17:24–25)

What happened to the poor family? They are being paid a 21 percent royalty on all petroleum production. Notice, following in-depth testing and a screenshot of the structure map produced by contour mapping for the deep gas and oil basin, the major oil company had high assurance that there is a major treasure hidden underneath. They only had to be hollowed out, brought to light, and put into production.

As we will realize from the subsequent pages, time and time again, the founders of different religions, saints, and the great ones—like Jesus the Christ, Saint Paul, Gautama the Buddha, Plotinus, Mohammed, Mahatma Gandhi, Meister Eckhart, Rabindranath Tagore, Ramakrishna, Abdu'l'Baha, Sri Aurobindo, and many others—have reported to us over and over once more about our "spiritual treasure" hidden in the field of our being. They have informed us that the unbounded intelligence and endless riches and resources are within us;

they themselves have excavated them. They themselves have verified them and supported them with wisdom and miracles after miracles. We see this in the sphere of Jesus. We see this by way of hundreds of others.

Universes Come and Go
You Will Always Be in Eternity

Advanced scientific knowledge of the fundamental laws of physics and their underlying connections has come to light in the last thirty years. To be exact, with high-precision measurement of the fundamental physical constants (contrasted with the mathematical constants, which have a fixed numerical value, but do not directly involve any high-precision physical measurements), physics, mathematics and computers, we reveal the presence and the glory of *God the Father* ('**1**') of everything that is unchanging and indestructible—the ultimate principle, the final law of the cosmos that is *above* the laws of nature. It embraces the foundation and all expressions of the universe, humanity, you, and your precious Holy Family unconditionally and absolutely (see appendix 2[1]). Specifically, it incorporates the ultimate principle, your living being—the Absolute Splendor who is an observable, outward expression and existence of the Holy of Holies—eternal, unchanging, and unmoving:

> I am the Alpha and the Omega, the One who is and who was and who is to come, the Almighty! (Rv 1:8)

Before the universe was created, the Alpha and the Omega—the Being, '**1**'—existed as timeless Holy Spirit who is called the Father of the Universe and of everything. Scientifically, in the *laws of physics*, it is represented as the equilibrium '=' sign. It is *the unchanging rest frame of nature.* Also, it is the magical place where all *transformations* take place and all *information* resides. Otherwise, it is the timeless, more comprehensive fusion of physics and *unchanging balance*; it is defined as the balance, symmetry, and stability in the laws of structure and organization and scale or else with the fundamental physical constants as the logarithmic *zero*. Not one thing was created without the '**1**'. The '**1**' is the *source* of life, and this life brings light to humanity (see appendix 2).

7

The law is *higher* and *greater* than the laws of nature. It is the *miracle zone* of the universe: *the point*, '**1**', the Sanskrit term *bindu*—which is *unmanifest*, unchanging, unmoving and Absolute, from which creation begins, where it is ultimately unified and from which we came. The *inverse* of the point is the *many* points (*zero point field*) or the Sanskrit term *maha bindu*—which is manifest, changing, moving, and relative. The unmanifest point and the manifest many points can be considered as the Being, "In him [the unchanging, *unmoving*, and indestructible] we live and move [hologram image projection] and have our being . . . We are in fact God's offspring [children]" (Acts 17:28–29).

Your Absolute Splendor—the beginning and end of all things— can be *recognized, known,* and *verified* with physical reality through the fundamental physical constants and the laws of physics. This basic truth, through verified science, will necessitate comprehensive awareness, knowledge, and updating of many important facts and volumes in physics, mathematics, cosmology, and religion.

My dearest heart, my dearest reader, the Absolute Splendor in human form, I love you compassionately as the Most High in your *manifest* human body, as the Christ in your personal universe. Infinite hugs of love and compassion to you, your lovely precious family, your dear friends, their fathers and mothers, sons, and daughters.

This work will help you enter into a higher life *and* universe by lending a hand to recall your *forgotten identity*—the Alpha and the Omega, who is, who was, and who will come. It is the awareness of the Most High and an *experience*: "I AM" in you (appendix 4). Also, this work will make it possible to *consciously link you* with the absolute life, Cosmic Consciousness, and source of pleasure, delight, understanding, holy joy, divine peace, and wisdom in God. Above all, it will assist you to become a conscious provider of unlimited symphonies of oneness and life in the absolute infinity—what your being, mind, hands, and legs are to you now.

At the present, you live and move and have your being in God. Your self is the transcendent being—the Absolute Splendor in discernible manifest human display. Our universe, earth, and the human race are only many appearances of the unchanging One as physical manifestation. Jesus said the kingdom of God (also everlasting life, Mk 10:17) is within you. There it is! You are looking at the physical manifestation of the Most High, the Alpha and the Omega God Almighty—who is, who was, and who is to come *in human expression*. So the kingdom is everywhere and in everything, and everything is in it. To be exact, your personal *interface* (your intellect, mind, and body) is not at all times consciously in synchronism with the oneness of the eternal world by way of Christ consciousness, health, rapture, bliss, and the unitive life (total engagement in the interests of the eternal). It is like your home electrical system not being connected to any electrical power grid or national complex or like your personal computer not being attached to the Internet and its global integrated network.

***

In his second inaugural address, President Abraham Lincoln illustrated the fighting between the North and the South with these words: "Both [North and South] read the same Bible, and pray to the same God; and each invokes His aid against the other . . . The prayers of both could not be answered . . . The Almighty has His own purposes."[2] Now consider the World War II and the last two-thousand-year planetary history. Essentially, what you experience depends on where you are or how far off you are of your personal equilibrium ('=') from the Absolute—total equilibrium ('**1**') (see appendix 2, the fundamental constant listing, or the logarithmic sliderule of physical relationships [LSPR]). The closer you are to the '**1**', the more you are in the zone and the depth of Christ consciousness. Namely, the light intensity changes depending on how far we are (personal or relative equilibrium) from the candle (absolute equilibrium). '**1**' is the *Mahabharata dharma* (right conduct, the basis of rightful living), your main focus to treat others as you treat yourself.

Following in the footsteps of the Council of Trent—the embodiment of the Counter-Reformation (1545 and 1563)—the First Vatican Council (1869–1870), Eastern traditions and a host of other ways of life, the Holy Scriptures (our spiritual treasures: Ancient Slavic, Ancient Iran, Ancient Mesopotamia and Urartu, Baha'i Faith, Buddhism, Chinese mythology, Christianity, the Church of Jesus Christ of Latter-day Saints, Manichaeism, Islam, Hinduism, Taoism, etc.) are written in allegorical interpretive method with *various levels of meaning* and secret language of cipher code that must be converted to translate or decipher the message and understand the meaning in the same spirit in which it was written. Codes are a way of changing a message (hiding or enciphering information) so the original meaning in religion, military, or personal secrets is camouflaged and protected.

For Bible students, I recommend reading Geoffrey Hodson's three-volume *The Hidden Wisdom in the Holy Bible*. For instance, Genesis utilizes "the tree of life in the middle of the garden and the tree of the knowledge of good and bad" (Gn 2:8–25) to encrypt the text. Similarly, Isaiah makes use of the "oil of gladness" (Is 61:3) and "water at the fountain of salvation" (Is 12:3) in a cryptogram. Jesus employs "parables" (Mt 7:6), "living water" (Jn 4:10), "the prodigal son" (Lk 15:11), and the like in secret code. Parables comprise about one third of Jesus's documented wisdom. Saint Paul employs baby or newborn "milk" and "solid food" in cipher code (1 Cor 3:2; Heb 5:12). The believers take milk, though it was time that they were ready for solid food (1 Pt 2:2).

Origen of Alexandria, also known as Origen Adamantius (184–253), was one of the leading Christian scholars, theologians, and ascetics in early Christianity. He is generally considered one of the most significant Christian theologians of all time. Origen was a very fruitful prolific writer with about two thousand dissertations in numerous branches of spirituality. He has been characterized as the greatest genius the early church ever produced. The churches of Arabia and Palestine regarded Origen as the final fundamental authority on all issues of theology. He created the *Hexapla*, the earliest essential version of the Hebrew Bible, which had the fundamental Hebrew copy as well as five distinctive Greek paraphrases of it. Saint Athanasius of Alexandria (also called Athanasius the Great [293–373]) and the three Cappadocian Fathers (Saint Basil the Great [330–379], Basil's younger brother Saint Gregory of Nyssa [335–395], and Saint Gregory of Nazianzus [329–389] were his most dedicated supporters. Here is how Origen describes the hidden wisdom in the Bible:

> What man of sense will agree with the statement that the first, second and third days in which the *evening* is named and the *morning*, were without sun, moon, and stars, and the first day without heaven? What man is found such an idiot as to suppose that God planted trees in Paradise, in

Eden, like a husbandman, and planted therein the tree of life, perceptible to the eyes and senses, which gave life to the eater thereof; and another tree which gave to the eater thereof knowledge of good and evil? I believe that every man must hold these things for images, under which the hidden sense lies concealed.[3]

And in Selecta in Psalmos, Patrologia Graeca XII, Origen continues:

The Holy Scripture is like a house in which all the rooms are locked, and the keys are not in the keyholes but scattered over the corridors and stairs; and none of the keys lying near the doors open those doors. The only way to interpret the Scriptures is therefore a close, methodical study of every text, every key. To find the right keys that will open the doors, that is the great and arduous task.[4]

Saint Jerome (342–420) was a Latin priest, theologian, and historian who translated most of the Bible into Latin (the translation that became known as the Vulgate) observes that:

The most difficult and obscure of the holy books contain as many secrets as they do words, concealing many things even under each word.[5]

Moses Maimonides, also known as Rambam (1135–1204) was a rabbi, scholar, physician, philosopher, and a prolific writer. His impact on the advancement of Judaism is immeasurable. In the post-Talmudic era, Maimonides was the most memorable notable in Judaism. He states this:

Every time that you find in our books a tale, the reality of which seems impossible, a story which is repugnant to both reason and common sense, then be sure that the tale contains a profound allegory veiling deeply mysterious truth; and the greater the absurdity of letter, the deeper the wisdom of the spirit.[6]

We find a similar line of thought in *Zohar*, the basic work of Jewish mysticism and greatest expositor of the Kabbalah:

11

Woe to the man who sees in Torah, [i.e., Law], only simple recitals and ordinary words! Because, if in truth it contains only these, we would even today be able to compose a Torah much more worthy of admiration. But it is not so. Each word of the Torah contains an elevated meaning and sublime mystery . . . The recitals of the Torah are the vestments of the Torah. Woe to him who takes this vestment for the Torah itself! The simple take notice of the garments or recitals of the Torah alone. They know no other thing. They see not that which is concealed under the vestment. The more instructed men do not pay attention to the vestment, but to the body which it envelops.[7]

It is unsafe to reveal the truth to the undeveloped, says Jesus:

Do not give what is holy to dogs or toss your pearls before swine. They will trample them underfoot, at best, and perhaps even tear you to shreds. To you the mystery of the reign of God has been confided. To the others outside it is all presented in parables, so that they will *look intently and not see, listen carefully and not understand* [my emphasis]. (Mt 7:6; Mk 4:11–12)

Countless individuals have benefitted in the *progressive revelation* of truth, freedom, peace, happiness, and agreement with reality from Abraham, Zoroaster (1500 BC), Moses (1400 BC), Krishna (1200 BC), Isaiah (740–686 BC), Lao Tzu (604–531 BC), Mahavira (599–527 BC), the Buddha (563–400 BC), Socrates (470–399 BC), Plato (427–347 BC), Aristotle (384–322 BC), Jesus (4 BC–30/33 AD), Muhammad (570–632), Baba Nanak (1469–1539), Baha'u'llah (1817–1892), and others (appendix 3). They have built a knowledge base about our Absolute Splendor and human advancement by way of *synchronizing* oneself to the unchanging Absolute who is the primal reality ('**1**'), the ultimate principle, and the *source* of all truth. They have provided paths to the spiritual awakening, which takes us to greater peace, prosperity, health, and bliss. They have taught us to look within ourselves to find God and advanced solutions for a better life and a more breathtaking universe.

Recently, we are blessed with additional truths that we can incorporate, validate, and corroborate with contemplation— in the vein of Leonardo da Vinci (1452–1519), Michelangelo (1475–1564), Nicolaus Copernicus (1473–1543), Galileo Galilei (1564–1642), Ralph Waldo Emerson (1803–1882), Henry David Thoreau (1817–1862), Nikola Tesla (1856–1943), Mahatma Gandhi (1869–1948), Sri Aurobindo (1872–1950), Ramana Maharshi (1879–1950), Albert Schweitzer (1875–1965), Pablo Casals (1876–1973), Teilhard de Chardin (1881–1855), Andres Segovia (1893–1987), Dag Hammarskjold (1905–1961), U Thant (1909–1974), Mother Teresa (1910–1997), Thomas Merton (1915–1968), Robert Muller (1923–2010), Martin Luther King Jr. (1929–1968), Sarah Chang (1980–), and numerous others.

Immeasurable happiness, delight, wonder, ecstasy, and joy are in store (the global transcendence of humanity) for us. Here, through *purification* and *training* of the self (learning and cultivating values, truth, nonresistance ['**1**', equilibrium, see LSPR in appendix 2], simplicity, and freedom), our opportunity to become more skilled at how to directly verify the Absolute consciousness in *ourselves* (appendix 1) —experience unity—live our *unmanifest* (the immutable) Absolute Splendor as *manifest* Christ in different forms and development (appendix 4) that are visible and invisible. Note that without the *purity* of heart, *stillness* of the mind, *fasting*, and *infused contemplation* the state of transcendence cannot be grasped by thought or speech. In the Book of Revelation *to* John by Jesus Christ, the unchanging truth or the Absolute Splendor symbolizes the oneness of existence:

> I am the Alpha and the Omega, the One who is and who
> was and who is to come, the Almighty! (Rv 1:8)

Psalm 85 also communicates the profound truth. It represents the Absolute capacity of the human race in the arena of everyday life, the Way, and thus stands for the One—formless, eternal, imperishable, unmanifest Most High, Brahman—the Absolute Splendor, *as* the manifest or awakened one—Christ, Ishvara, Mahavira, or the Buddha, in the relative world:

> They know not, neither do they understand; they go about
> in darkness; all the foundations of the earth are shaken. I

said: You are gods; all of you are sons of the Most High; yet like men you shall die, and fall like any prince. Rise, O God; judge the earth, for yours are all the nations. (Ps 82:5–8)

The *ultimate principle*, '**1**', is observable in the laws of physics (appendix 2) as the unmoving, unchanging *stillness* ("Be still and know that I am God," Ps 46:10). In the equations of physics, the final law of nature is evident as the transformation interface and is marked with an equal sign, '='. In cosmology, the ultimate principle is evident across our cosmos as the unmoving, impenetrable *wall* or the barrier that, no matter how powerful our telescopes become, we will not be able to violate this boundary to space and time.

Furthermore, the barrier separates elements, physical quantities, electrons, protons, and neutrons. It is also the *empty* interval between two musical notes. Note, within the ultimate principle, a never-ending number of other universes (multiverse) live and move and have their beings as our own does. In addition, the same torrent of life flows throughout the universes.

Mevlana Jalaluddin Rumi (1207–1273), who is commonly regarded as one of literature's greatest mystic poets and who realized that we are on the journey to your essence, noted, "Hundreds of universes fit in the eye of a needle."[8] As we grow in the unity of life—purity of intent, forgiveness, social welfare, inner peace, knowledge, simplicity, detachment, holiness, compassion, truth, and self-control—we will not only be able to visit various parts of our universe but also our next-door universes, which *interpenetrate* our universe. We will also see that all these universes, population, and life constitute just *one point of awareness*, which is the Absolute Splendor (see the parable of the mustard seed, Mt 13.31, Mk 4:30, and Lk 13:18).

Experience God within Yourself
Invest in You

How can one achieve inner peace, the peace of God, that passeth all understanding? How can one acquire firsthand knowledge or directly experience the God within ourselves and within everyone? Aldous Huxley once said that Western religion attempts to know God, whereas

Eastern religion attempts to be God. All the answers you search for are within you, within *each point* of the Absolute Light of Certainty. The *Bhagavad-Gita* (written as early as the fifth-century BC or as late as the first-century AD) exemplifies the heart of the most holy Hindu Scripture, the Vedas (are universally recognized as the earliest spiritual revelations in the history of humanity). This insight can be found:

> Whatever you wish to see can be seen all at once in this [human] body. This universal form can show you all that you now desire, as well as whatever you may desire in the future. Everything is here (at every point) completely. But you cannot see Me ['**1**', the Clear Light of the Void, the Realm of infinite possibilities and miracles] with your present eyes. Therefore I give to you divine eyes [the Cosmic State of Consciousness] by which you can behold My mystic opulence. (11:7–8)[8]

You can attain direct knowledge and experience the Absolute Splendor within yourself by purifying your stream of consciousness from contamination and "toxic waste" by eating of the tree of knowledge of good and bad. You can journey from unconscious divinity to conscious divinity.

There were times in our growth of the understanding (before the awakening, Cosmic Consciousness, or blissful enlightenment) where we were constrained from knowing our identity, from accessing the fundamentals of nature, or from recognizing our hallowed divinity. In the Old Testament (Gn 2:15), the Lord God (celestial personage or angelic being) gave the human race this command:

> You are free to eat from any of the trees [parts] of the garden [human body], except the tree of the knowledge of good and bad. From that tree you shall not eat; the moment you eat from it, you are surely doomed to die.

Additionally, we also learn (Gn 3:4–6 and 3:22) that

> you certainly will not die! No. God knows well that the moment you eat of it, you will be like gods who know what is good and what is bad . . . [And] See! The man has become

like one of us [gods] knowing what is good and what is bad! Therefore, he must not be allowed to put his hand to take fruit from the tree of life also, and thus live forever.

In sermon 26, "the Holy Spirit, like a rapid river, divinizes us," maintains Eckhart:

> There is a river whose streams refresh the city of God, and sanctifies the dwelling of the Most High.[9] The rapid or quick flowing river has caused the city of God to rejoice. (Ps 46:4)

From the Revelation, we learn that through living *in one Spirit* (pure heart, righteousness, compassion, stillness of the mind, and "thy will be done" setting), we can unequivocally access the Absolute Splendor (Eph 2:18) and come to *share in the divine nature* (2 Pt 1:4; Col 1:15). To those who have won the victory over wrongdoing,

> I will see to it that the victor eats from the tree of life which grows in the garden of God. (Rv 2:7)

The garden of God denotes the human body. Therefore, how can one realize firsthand knowledge and experience our Absolute Splendor? Answer: with absolute *commitment to eat the fruit of the tree of life and knowledge,* which is the kingdom of God within you. Can we elaborate on the fruit of the tree of life and knowledge? The Old Testament and the New Testament characterize this fruit ("streams of living water") as the *Holy Spirit* (Jn 7:37–9), *hidden manna, bread of life,* oil of gladness (Is 61:3; Heb 1:9), the *fruit from the tree of life,* the *bubbling spring* where you become drunk (the Gospel according to Thomas), the oil of gladness (Ps 45:7), and the *living water* (Jn 4:14, 7:38):

> Your throne, O God, stands forever and ever; tempered rod is your royal scepter. You love justice and hate wickedness; therefore God, your God, has *anointed you with the oil of gladness* (my emphasis) above your fellow kings. With myrrh and aloes and cassia your robes are fragrant; from ivory palaces string music brings you joy. (Ps 45:7–9)

as well as

You have loved justice and hated wickedness, therefore God has *anointed you with the oil of gladness* [emphasis my] above your fellow kings. (Heb 1:9)

and

But whoever drinks the water I give him will never be thirsty; no the water I give shall become a fountain within him leaping up to provide eternal life. (Jn 4:14)

and also

If anyone thirsts, let him come to me; let him drink who believes in me, Scripture has it: "From within him rivers of *living water shall flow* (my emphasis)." (Jn 7:38)

In the Eastern philosophy and religion, this fruit is known as *kundalini*, the serpent power; it is the creative force of the universe, which is coiled at the base of the spine.

Why do economies rise and fall? How can we determine the spectrum of physical quantities? How can one compose Bach Brandenburg Concertos or Mozart Symphony No. 40 in G minor, K. 550? How can we generate the laws of physics with a computer? How do ideas spread? What are the various states of consciousness beyond that of everyday experience? How do we rise from self-centered desire for power, status, wealth, and control to love and compassion for all? How do we transcend human consciousness? These and other engaging challenges are accessible within the Absolute Splendor with *infused contemplation*. This knowledge is on hand, and we discuss it in this work.

Similar to the Vedic *Bhagavad-Gita* reference earlier, "all the solutions and answers you search for are inside you," says Saint Symeon the New Theologian. Saint Symeon (949–1022) was a Byzantine Christian, renowned monk, ascetic, poet, and the abbot of the Monastery of Saint Mammas for twenty-five years. He is venerated by the Catholic Church and the Orthodox Church for his reputation for sanctity and his teachings and who spoke from personal direct access to the Godhead. He put it this way,

Search inside yourself with your intellect so as to find the place of the heart, where all the powers of the soul reside. To start with, you will find darkness and an impenetrable density. Later, when you persist and practice this task day and night, you will find, as though miraculously, an unceasing joy. For, as soon as the intellect attains the place of the heart, at once it seems things of which it previously knew nothing. It sees the open space within the heart and it beholds itself entirely luminous and full of discrimination. From then on, from whatever side a distractive thought may appear, before it has come to completion and assumed a form the intellect immediately drives it away and destroys it with the invocation of Jesus Christ . . . . The rest you will learn for yourself, with God's help, by keeping guard over your intellect and by retaining Jesus in your heart.[10]

From the verified experiences of mystical tradition—Judaism, Christianity, Islam, Hinduism, Buddhism, Taoism, et cetera—it is obvious that within the essentials into the depths of the heart to discover the hidden treasure of the inner kingdom, the path of God-realization (the ground that is the source of all becoming and the pure knowledge that the Absolute is in us, in all things, and all in the Absolute) is *equivalent.*

In summary, all through the centuries, numerous human beings (appendix 3) embarked on the awakening, purity of mind, and training of the self in the ascent of the path that directs to the blessedness of the unitive life. Many of them had direct experiential encounter with the Absolute Splendor within them. With experiential encounters, additional talents, abilities, and powers accompanied them—including an awareness of oneness, knowledge, wisdom, healing, visions, revelations, perfume-like fragrance from their bodies, walking on fire or water, bilocation, resurrecting the dead, and so forth.

As more people believed the revealer's message or reproduced these talents within themselves, religions formed. Religions of Judaism, Hinduism, Taoism, Christianity, Islam, Jainism, and the like now are communicating *some* of these realities. Through hearing the message of liberation from religions, schools, corporations, or governments, the whole world in time can believe it. By believing it, we may hope; and by hoping for it, we will find peace, love, righteousness, compassion, and the Absolute Splendor within ourselves.

In this work, in addition to the Cosmic Consciousness, we bring divine revelation of oneness through the exact scientific evidence. The rigorous scientific proof encompasses the laws of physics and the fundamental physical constants, which is built into the universe (appendix 2). Specifically, every law of physics has inner unity and is enduring witness to the innermost being of the Absolute Splendor in created realities (Jn 1:1–18). In effect, to see the creation is to observe the Absolute Splendor face-to-face; to see Jesus is to observe the Most High in human form (Jn 5:36–40).

What we witness through the laws of physics and the fundamental physical constants is this: *the Most High is in every physical equation and in each fundamental physical constant.* An equation of time, space, gravity, motion, and so forth cannot exist without the equal sign symbol ('=') and the Absolute Equilibrium ('**1**').

Max Planck observed with his Planck postulate (the electromagnetic energy could be emitted only in quantized form) and his effort with the opposition attitudes in his former days: "A new scientific truth does not triumph by convincing its opponents and making them see the light, but rather because its opponents eventually die, and a new generation grows up that is familiar with it."[11]

Therefore, the beginning and end of all things can be known with certainty from created reality by the light of human reason (Rom 1:20). Hence, in the Epistle of Paul to the Galatians, he states, "The life I live now is not my own; Christ is living in me" (Gal 2:20). We are on the outing (pleasure trip, expedition) to our fundamental nature.

God Cannot Be Known Except by Himself
God Knows Himself

Dearest reader, give up all uncertainty except one: Who am I? The seeker is he/she who is in search of himself/herself. At the very beginning of his public life in the Gospels according to Matthew and Luke, Jesus declared this:

The kingdom of heaven is at hand. [Also,] the reign of God is already in your midst. (Mt 4:17; Lk 17:21)

The reign of God is like riches buried in the field, resembling the Kentucky oil treasure hidden in the ground, which had to be brought to light and developed. We should also cultivate our immeasurable treasure: the Absolute Splendor. As pointed out previously from the demonstrated knowledge of *mystical tradition*—Christianity, Islam, Hinduism, Buddhism, Judaism, Jainism, Baha'i, and so on—the central trail of spiritual awakening and God-realization is equivalent—the same.

There are thousands of excellent spiritual-awakening and God-realization Holy Scripture, meditation and prayer books, first-rate conferences, and marvelous discussions. In essence, one can study a map without engaging in traveling.

***

Saint Thomas Aquinas (1225–1274) was a Dominican friar, Catholic priest, doctor of the church, and truth seeker. He is regarded as one of the Catholic Church's supreme theologians, thinkers, and philosophers. The English philosopher Anthony Kenny—whose research has been on philosophy of religion, the philosophy of Wittgenstein, and on the philosophy of the mind—regarded Aquinas as one of the dozen greatest philosophers of the Western world. Aquinas has written commentaries on scripture and Aristotle, *Disputed Questions on Truth*, *the Summa contra Gentiles*, and the unfinished work *Summa Theologica* (*Summa* for short), known as one of the classics of the history of philosophy and one of the most significant works of Western writing. *Summa* deals with the existence of God: the Trinity, the Creation, the angels, man, the divine government, the last end, human acts, habits, law, grace, faith, active and contemplative life, the states of life, the Incarnation, the sacraments, the Resurrection, and the Last Things. It is intended for theology students and the well-read laity. Shortly, we will take a look at the Gospel according to Thomas, the saying of Jesus: "Let him who seeks, not cease seeking until he finds and when he finds, *he will be troubled* [my highlight]."[1]

Near the last part of his life, Saint Aquinas was celebrating a morning mass in infused contemplation. Following this event, in the vein of the Gospel according to Thomas, being troubled, Aquinas was markedly disturbed and stopped writing. When asked why the saint stopped writing, Saint Aquinas replied, "I can do no more; such things have been revealed to me that all that I have written seems to me as straw. Now, I await the end of my life."[2] Subsequently, Saint Aquinas refused to put pen to paper or continue his *Summa*. This event made its way into the canonization process for Saint Aquinas.

Note 1: *Doctor of the church* (i.e., Augustine of Hippo, Thomas Aquinas, Hildegard of Bingen, Pope Gregory I, Teresa of Avila, etc.) is a designation given by the Catholic Church to saints known as having made a major contribution to Church doctrine.

Note 2: *Meditation* or *stillness of the mind* in religious traditions has been employed for inner peace, enlightenment, and self-realization (see attachment 1). Meditation reduces pain, anxiety, depression, and stress. It increases peace, *awareness on self-concept*, and health (cardiovascular, neurological, and psychological). Meditation or stillness of the mind rebuilds the brain's gray matter associated with memory, sense of self, empathy, and stress. It can also be used to *mine your own being*—mind and will on art, business, politics, religion, science, and so forth for insights on wisdom, truth, and forward thinking of anticipation of the future to improve strategies and decisions.

You hollow out, unearth, or mine your mind, awareness, and experience—including what Aquinas could have read or cited (i.e., Christian sacred scriptures, Aristotle, Augustine of Hippo, Avicenna, Averroes, Al-Ghazali, Boethius, John of Damascus, Paul the Apostle, Dionysius the Areopagite, Maimonides, Anselm, Plato, Cicero, and Eriugena) when he produced his works.

Mining the Eternal Being
A Measure of Knowledge beyond the Intellect

Contemplation is usually connected with belief, prayer, meditation, or *thinking*, reflecting and deliberating about something in the religious terms or life. Customarily *we* reflect on, *we* think about, and *we* mull over something. In mining the Eternal Being, we take ourselves out of the picture—that is, the *input* in infused contemplation is the *Eternal Being*, not us. In Matthew, Jesus stated, "Your Father knows what you need before you ask him" (Mt 6:8).

In Christianity or Christian mysticism, *infused* contemplation is considered an *unearned* or *unmerited* gift, a *supernatural gift*, by which mind and heart together will become wholly centered in the Eternal. Consequently, the soul cannot contemplate wherever or whenever it wants, but only when God desires and in the measure and degree He wishes. Under this conviction, the intellect receives insights into *spiritual truths*, and the will is alive with divine love.

Here, in *Your Miracle after Miracle Life*, we consider a wider and broader reading than the patristic tradition (the tradition of the church incorporating the scripture and the teachings of the fathers). We suggest that God does not show favoritism (Rom 2:11). That infused contemplation is not *unmerited* bequest or a supernatural gift.

In our land of infused contemplation, there are *no favors* from the Eternal! No gifts. *You earn it*: "Blest are the single-hearted [pure in heart], for they shall see God" (Mt 5:8) and "I will see to it that the victor [over wrongdoing] eats from the tree of life which grows in the garden of God" (Rv 2:7) and also "To the *victor* [my emphasis], I will give the hidden manna" (Rv 2:17).

Infused (or higher) *contemplation* suggests *mining the Eternal Being* whenever you need a greater penetration of truth—the Absolute knowledge and forward thinking—through a range of possible actions that, like the laws of physics, are unchanging, infinite, and everlasting. For example, the search for the fundamental laws of nature or understanding unification of the visible reality of nature has been one of the major challenges of philosophy, science, mathematics, physics, and religion. It confronts both human logic and our human mental powers.

On the back cover of *Your Miracle after Miracle Life*, we reference Max Planck from Berlin University, Eugene Wigner from Princeton University, and Steven Weinberg from the University of Texas. They have addressed the ultimate mystery of nature, the full meaning of life, and the possession of the final physical principles. Planck, Wigner, and

Weinberg maintain that these challenges are beyond human reach, viability, and unsolvable now.

In appendix 2, you will find 1993 refereed scientific paper from the Institute of Mathematics of National Academy of Sciences in Kiev, Ukraine. The paper shows and verifies scientifically the ultimate unifying principle (the Absolute, the Unchanging Eternal) and the ultimate mystery of nature. In appendix 1, you will also find how to see God face-to-face through a union with the Eternal. At this juncture, you realize that all knowledge that was, is, or ever will be is here and now at the Eternal Being. Similar to a radio or a TV program, in infused contemplation, one can fine-tune into them, as the *Bhagavad-Gita* and Saint Symeon earlier presented.

To enter or go into infused (sometimes called intuitive) contemplation requires *synchronizing* yourself with the Eternal, the mind of God (the one Self, Atman, indivisible consciousness), which the Epistle of Paul the Apostle to the Philippians expresses, "Then God's own peace, which is beyond all understanding, will stand guard over your hearts and minds, in Christ Jesus" (Phil 4:7).

Note, in Revelation to John, Jesus Christ stated that

> I am the Alpha and the Omega, the One who is and who was and who is to come, the Almighty! (Rv 1:8)

Additionally, when we study universal truths about nature, life, chemical elements, the laws of physics, and the fundamental physical constants, we encounter consistency, completeness, and connectivity to a common starting point that is unmoving and unchanging—which is so rigid and so severe that it cannot be distorted by creating logical absurdities. Also, when one integrates these attributes of God into one depiction, we arrive at a holographic, multidimensional *equilibrium point*, which can be observed, modulated, and accessed from different angles by different observers, which Matthew (13:31–32), Mark (4:30–32), and Luke describe in the parable of the mustard seed:

> Then he said: "What does the reign of God resemble? To what shall I liken it? It is like mustard seed [the smallest of all seeds], which a man took and planted in his garden. It

grew and became a large shrub and birds of the air nested in its branches." (Lk 13:18–19)

Reaching it is entering the kingdom of heaven and being born again: "Flesh begets flesh, Spirit begets spirit. Do not be surprised that I tell you, you must all be begotten from above" (Jn 3:6–7). Then by way of *spiritual union* with the Eternal or *Nirvana* (appendix 1, item 11), "your attitude must be that of Christ."

By putting on the mind of Christ (Phil 2:1–11), you mine the Eternal by way of infused contemplation. Only in deep peace and stillness of the mind (beyond feeling of a body or surroundings) is the unchanging truth found. Only when the lake is completely still will it reflect the stars clearly. Similarly, only when the mind is perfectly still will it reflect the nature of the self.

Christian mystics refer to union with the Eternal Light as "beatific vision," by Buddhists "Nirvana" (unbrokenness, oneness), by Hindus "Moksha," and by Dr. R. M. Bucke "Cosmic Consciousness."

In union with God, Nirvana, or beatific vision *state*, one listens to *the voice of the* Eternal Silence. In that heavenly peace, light, and stillness, one can also experience beautiful color images, delicate flower perfume or fragrance; melodious divine music; or subtle sound, the universal pulse of life, or the flowing stream of consciousness (Sanskrit, *nada*), and so on.

On one occasion, while planting table grape plants in the backyard garden, while I was fasting, I heard billions upon billions of instrument orchestra playing together. Every galaxy, star, our sun, DNA (double helix carrying genetic instructions for the development, growth, functioning, and reproduction of organism), electron, proton, atom, and the like were producing extraterrestrial music.

Wolfgang Amadeus Mozart (1756-1791) created more than 600 works, many which are recognized as summits of symphonic concertainte (containing one or more solo parts), operatic, chamber and choral music. In the vein of Ludwig Beethoven along with Johannn Sebastian Bach who could through infused contemplation access the 3-D Absolute music, Mozart is regarded as among the greatest classical composers of all time. "Father of the Symphony" and "Father of the

String Quartet" Joseph Haydn stated: "Posterity will not see such a talent again in 100 years." In a letter Mozart wrote:

> Provided I am not disturbed, my subject enlarges itself, becomes methodized and defined, and the whole, thought it be long, stands almost finished and complete in my mind, so that I can survey it, like a fine picture or a beautiful statue, at a glance. Nor do I hear in my imagination the parts successively, but I hear them, as it were, all at once.

Thinking, reflecting, and deliberating obliterates spiritual union in the heavenly state. In the celestial state, the intellect can receive deep insights into the foundation of nature, righteousness, and common types of forward thinking. Aspects of infused contemplation are discussed further.

Saint Teresa of Avila reveals:

> It was granted me to perceive in one instant how all things are seen and contained in God. I did not perceive them in their proper form, and nevertheless, the view I had of them was of a sovereign clearness and has remained vividly, impressed upon.[1]

In 1539, Saint Ignatius of Loyola (1491–1556), Peter Faber, and Francis Xavier formed the Society of Jesus (Jesuit order) and developed a simple set of meditations, prayers, and mental exercises. Saint Ignatius, the great teacher of meditation, recognized that "a single hour of meditation . . . had taught him more truths about heavenly things than all the teachings of all the doctors [of the Church] put together."[2]

In "Union without Ceasing" (appendix 1), published in 1974—which is a part of *Yes It's Love: Your Life Can Be a Miracle*, a spiritual awakening book—I found assistance from Jesus the Christ, Mother of Perpetual Help Mary, Saint Michael the Archangel, and Maitreya the World Teacher. Additionally, I had help from Saint Anthony of Padua, the apostles and saints, scholars, and seers' guidance in achieving the union with God.

Harmony and unity without ceasing with God is attained via inner peace, purity of heart, and stillness of the mind (absence of thought or no thought) by way of meditation, periodic fasting, silence, compassion, and detachment.

Comparing my living to the immense cosmic silence of mountain monasteries, monks on Mount Athos, or tranquility and serenity of the hermits of the desert, bringing to fruition the "Union without Ceasing" with the Absolute was challenging on many levels.

Fifteen years earlier, I had coped with the flashbacks of the horrors of War World II, Soviet and Nazi violence, and British (night) and US (day) bombardments of areas inhabited by civilian population in Ukraine, Poland, Czechoslovakia, Hungary, and Austria. In Revuca, Czechoslovakia, Russian air-landing units (paratroopers, or Locust Warriors, as they were called) dropped like bombs on top of us, behind the German lines. These airborne operations were used to seize airfields and city centers in advance of the land forces. The Russian parachute forces were not taking prisoners.

There were eighteen children and adults in our group who escaped the same day of air-landing, from Czechoslovakia to Hungary, where we thought the Germans were. Not to be observed by the border patrol or the paratroopers, we walked in the mountains for five days at night. We had enough food for two days. The next two days, we had no food and no water on the fifth day. My mother, who was pregnant, lay on the ground and said, "I have no strength to go on. Please leave me here."

In the Sátoraljaújhely Jewish ghetto in Hungary, our family of four (Christians) had our sleeping facilities on the synagogue floor with beautiful windows. During the bombings, Hungarian soldiers, who guarded the ghetto, would gather us into WWI military frontline dugouts to watch the terror campaign and fireworks at night. In Austria, the Strasshof Concentration Camp is where we were transported from the Sátoraljaújhely ghetto by the Nazi in a Holocaust boxcar train and escaped at the Strasshof railroad yard before entering the concentration camp complex.

In Vienna and Salzburg, there were the customary heavy bombings, with bombs, planes, and pilots coming down in no time. I watched the violence over Vienna from a chicken coop roof where we were hiding from the Nazis. A Samaritan woman allowed us to stay there for four days and brought us food to eat. We had no documents. My father hiked to Vienna and obtained ID and traveling papers.

Hungarian Jews had an advantage over other prisoners in Strasshof Concentration Camp. SS *Obersturmbannführer* Otto Adolf Eichmann (who managed the deportation of Jews to extermination camps and was captured by the Mossad in Argentina in 1960 and found guilty of war crimes in Jerusalem) agreed with the leaders of the Relief and Rescue Committee of Budapest to spare "one million" Hungarian Jews in exchange of money, trucks, and other military goods. Although five million Swiss francs were paid to the SS in exchange for 21,000 Jews sent to Strasshof, the arrangement eventually failed. However, virtually all the Jews at Strasshof lived to tell the story.

Note: Auschwitz Concentration Camp was a complex of over forty concentration and extermination camps created and operated by Nazi Germany in subjugated Poland. It was functional throughout World War II, where 1.3 million individuals were sent from all over German-occupied Europe, and 1.1 million died of gassing, execution, starvation, disease, or beatings. At least eight hundred prisoners tried to escape, 144 successfully. On August 14, 1941, when a camp prisoner seemed to have escaped, SS Second-in-Command Karl Fritzsch ordered that ten other prisoners should die by starvation. Franciszek Gajowniczek (prisoner 5659), a Roman Catholic born in Poland, was selected at roll call, cried out in agony that he had a wife and two children. Maximilian Kolbe (1894–1941), a Conventual Franciscan friar (prisoner 16670), stepped forward and said, "I am a Catholic priest from Poland; I would like to take his place because he has a wife and children." The exchange was allowed, and the reprimand carried out.

To my knowledge, there were no reprisals at Strasshof Concentration Camp on our escape. Maximilian Kolbe was canonized by his fellow Pole John Paul II on October 10, 1982, and Francis Gajowniczek was present for that formal procedure.

A light heart is eternal. "There is no greatest love than this: to lay down one's life for one's friends" (Jn 15:13), said Jesus. Similar to the Saint from Auschwitz, Kolbe, the Lord Jesus knew about his impending suffering from the elders, the chief priests, the teachers of the Law of Moses, death, and the third day raising to life (Mt 16:21), also would have known *before he was born on earth* (Is 53:3–7, Ps 22:4–18, Jn

1:10–14). Kolbe was seeking to help the Gajowniczek family. Jesus was aiming to help us, humanity, to jog our hearts and minds by reminding us (the prodigal son) concerning our identity and our divinity. Please see notes.[3]

Yahshua (Jesus), his disciples, and the villages of Nazareth and Capernaum in Galilee where Jesus spent most of his life, were Aramaic-speaking people. Jesus talked and preached in Aramaic, not in Greek, Latin, or in English. As stated earlier, the Gospels most likely were at first recorded in Aramaic, which then were translated into Greek, Latin, and English.

The World Is the Way It Is

Before the crucifixion and during the crucifixion, Jesus was continually aware, awake, and *conscious* of His and *our* being along with His and our divinity. On a number of times, Jesus affirmed these: "The Father and I are one" (Jn 10:30), "You are gods" (Jn 10:34), and "I am in my Father, and you in me, and I in you" (Jn 14:20). "Before the world was created the Word *already* existed: he was with God, and he was the *same* [my accent] as God" (Jn 1:1–2). The Word, Christ, was with God, and the Word was God.

Additionally, why did Jesus come to earth, and what sequence of events were ahead of Him. In Mark, Jesus stated:

> The Son of Man had to suffer much, be rejected by the elders, the chief priests, and the scribes [of the Law], *be put to death, and rise three days later* [my accent]. He said these things quite openly. (Mk 8:31–32)

a) What was the location of Jesus's crucifixion and death? Answer: Golgotha.

b) Where did Jesus start his crossing to Golgotha? Answer: Jesus began his journey to Golgotha from the Sanhedrin trial of Jesus (which acted as the highest court of justice and the Supreme Council in ancient Jerusalem), south of the Altar. The Synoptic Gospels (Mt 26:19–50; Mk 14:12–46; Lk 22:7–54;

Jn 13:21–30) state that after having the Passover meal (the first day of the Feast of the Unleavened Bread) with his disciples, Jesus was arrested later that night, the high of the first day of the Passover. That night, he was accused of many things and *condemned to death* because he claimed to be the Messiah, the Son of God, violating the Sabbath law (healing on the Sabbath), etc. Very early in the morning, the elders, the chief priests, the teachers of the Law, and the Council of Sanhedrin put Jesus in chains and handed him over to Pilate for his execution (Mk 15:1).

c) When did the chief priests with the elders, the teachers of the Law, and the whole Sanhedrin bring Jesus to Pilate? Answer: The Synoptic Gospels state, "Very early in the morning" (Mk 15:1). The Sanhedrin mock trial that resulted in the crucifixion of Jesus Christ took place *illegally* through the night. We estimate Jesus was arrested 2:00–3:00 a.m. The Sanhedrin governed the criminal law and had self-governing authority of police and right to make arrests through its own officers of justice. It was also authorized to judge cases that did not entail the death penalty; only capital cases involved the approval of the procurator. To get a speedy confirmation from the Pilate and not to instigate population to oppose their decision, Jesus was brought in as soon as it was daylight, just after sunrise, as the trials in the forum in Rome began.

d) Who was the Roman prefect of Judaea at the time of Jesus's death? Answer: Pontius Pilate served as the Roman prefect of Judaea between AD 26 and 36. He was the fifth governor of the Roman province of Judaea, serving under Emperor Tiberius. The Jewish philosopher Philo of Alexandria (AD 50) characterized the prefect for his cruelty, bribes, and executions without trial, insults, and robberies.

e) When did the Stations of the Cross start? Answer: the real proceedings that are the stations started when Jesus was sentenced to death by Pontius Pilate. The tradition of commemorating the Stations of the Cross started following Jesus's death, when people started walking his crossing to Golgotha. Signs were set up in Jerusalem at the locations where the stations took place.

f) What were the places and the time and distance elements Jesus had to deal with between His arrest and Golgotha? Answer: first-century Jerusalem measured about 1,700 meters (1,609 meters per mile) south to north and about 1,000 meters east to west. The crucifixion ordeal (arrest 2:00–3:00 a.m. to crucifixion 9:00 a.m.) entailed six to seven hours. All the walking between Jesus's arrest and crucifixion, as described in the Gospels, might have taken 1.5– 2 hours as well as 4–5 km:

Arrested at Garden of Gethsemane: 2:00–3:00 a.m. (Mk 14:37; 14:41).

Walked to the house of the Ex High Priest, Annas.

Taken to the High Priest Caiaphas (Mt 26:57; Mk 14:55) and the Sanhedrin at night (as a rule, trials were allowable in daylight and *not* at night or on the Sabbath).

Questioning by Pontius Pilate (Lk 21:1).

Interview by Herod Antipas (Lk 23:7).

Back to Pilate (Lk 23:11).

Pilate handed Jesus over to the Roman soldiers to be brutalized. He knew Jesus was innocent. Pilate washed his hands of the crucifixion affair. The Shroud of Turin (the ticking hydrogen time bomb *with* Sudarium of Oviedo stained in blood companion to the shroud, measuring 84 x 53 cm [33 x 21 inches], the handkerchief or napkin, which touched Jesus according to the Gospel of John [Jn 20:6–7], that had been around His head was now in the Cámara Santa in the Cathedral of San Salvador, Oviedo, Spain) showed blood of painful suffering. From the Shroud of Turin one can see that Jesus was beaten and tortured almost beyond human recognition. There are bloodstains from wounds on the head, back, side, arms, legs, and feet. In the Roman Empire, whips with small pieces of metal or bone at the tips were usually utilized as a prologue to crucifixion. According to the Gospel accounts, this occurred prior to the crucifixion of Jesus Christ.

Crucified at Golgotha (outside the city walls, probably outside the northern wall and several hundred meters from the temple and Pilate palace, which was build by Herod Antipas's father Herod the Great).

It was nine o'clock in the morning when Jesus Christ was nailed to the cross through His anatomic *wrists*, which is part of the hand (Jn 20:27, Lk 24:39–40)—and not the middle of His palms, which could have easily been torn apart—as the Shroud of Turin (the 437 cm [about 14.3 feet long] by 111 cm [about 3.6 feet wide] burial linen cloth of Jesus) shows. Like an x-ray image, bleeding at the wrists just above the palm burned into the cloth by means of radiation created by His Resurrection. The pain in this area is unbearable because of a hypersensitive nerve that bypasses this spot.

The gamma ray image on the burial cloth of Jesus (which is beyond the visible spectrum, ultraviolet rays, and x-rays) of the shroud would require billions of watts of radiation to turn burial linen into perfectly photographically sensitive material, which surpasses the production of any supply of radiation now intended for that size linen. In the Hiroshima and Nagasaki atomic bomb detonation, one can see human gamma-ray images on building walls. The Shroud of Turin image must have been produced *within* the burial linen cloth and not outside the Shroud of Turin. The handkerchief or napkin that touched Jesus's head was *not* subjected to the Resurrection gamma-ray radiation. It will be very useful in future research.

In Luke (23), John (19), Matthew (27), and Mark (15), to the last breath, we witness Jesus being *conscious*, mindful, and cognizant:

> Two others who were criminals were led along with him to be crucified. When they came to the Skull Place (Golgotha), as it was called, they crucified him there and the criminals as well, one on his right and the other on his left. Jesus said, "Father, forgive them; they do not know what they are doing." (Lk 23:32–34)

> One of the criminals hanging in crucifixion blasphemed him! "Aren't you the Messiah? Then save yourself and us!" But the other one rebuked him: "Have you no fear of God, seeing you are under the same sentence? We deserve it, after all. We are only paying the price for what we've done, but this man has done nothing wrong." He then said, "Jesus, remember me when you enter upon your reign." And Jesus replied, "I assure you: this day you will be with me in paradise." (Lk 23:39–43)

"Near the Cross of Jesus there stood his mother, his mother's sister, Mary the wife of Clopas, and Mary Magdalene. Seeing his mother there with the disciple [Saint John] whom he loved, Jesus said to his mother:

> "Woman, there is your son." In turn he said to the disciple, "There is your mother." From that hour onward, the disciple took her into his care. (Jn 19:25–27)

After that, Jesus realized that everything was now "finished [completed]" (Jn 19:28). Persian poet, Islamic scholar, and Sufi mystic Jalaluddin Rumi, who lived about twelve centuries after Jesus, declared this:

> It is as if a king had sent you to a country to carry out one special, specific task. You go to the country and you perform a hundred other tasks, but if you have not performed the task you were sent for, it is as if you have done nothing at all. So people have come into the world for a particular tasks, and that is our purpose. If we don't perform it, we will have done nothing.[4]

Our personal particularized task here on earth is to consciously realize our elapsed, forgotten divinity, to become conscious that God is in us and we are in God. That "God the Logos became what we are, in order that we may become what He Himself is," as Saint Irenaeus stated earlier. Furthermore, that "you are the light of the world" (Mt 5:14), and remember that "there is no greater love than this: to lay down one's life for one's friends" (Jn 15:13), as Jesus Christ did for us to save and free the lost.

The book of Timothy says, "Christ came into the world to save sinners" (1 Tm 1:15). To save the lost (Mt 18:11) and to bring back to life the dead. We quote Revelations, "Keep firmly in mind the heights from which you have fallen" (Rv 2:5). He used the lost sheep (Lk 15:1–7), the lost coin (Lk 15:8–10), and the lost son (Lk 15:11–32) parables for the entrapped within mortal bodies and the knowledge and technology necessary for the escape to remind us about our identity: that the Kingdom of God is within us and that Exodus 3.14 the "I AM" and the "I AM WHO I AM" ["I Am that I Am"] is in us.

In my infused contemplation, hanging on the cross, Jesus was filled with very, very great joy of wholeness, profound happiness, and ecstatic exultation that He *has succeeded* in His specific task and that He will come to life (raise from the dead) in three days. Not only the Resurrection but He also knew that "all authority in heaven and on earth has been given to me. Therefore go and make disciples of all nations, baptizing them in the name of the Father and of the Son and of the Holy Spirit . . . And surely I am with you always, to the very end of the age" (Mt 28:16–20; Mk 16:14–20; Lk 24:35–53; Jn 20:19–21, 25).

"Jesus cried out [in Aramaic] with a loud voice and said, *'Eli, Eli, lama shabachthani'"* (Mt 27:46), or in Mark's version, *"Eloi, Eloi, lama shabachthani"* (Mk 15:34). When translated without intermediary translations, from the Eastern Text Bible (Lamsa Bible, Mt 27:46) to English of the New Testament, it means, "My God, my God, for this I was spared!" "This was my destiny" or "My God, my God, for this I was born!" That was my assignment; that was my providence.[5] Thank you, Father! Thank you, Father!

Language of Jesus. As stated earlier, it is commonly agreed that Jesus, his disciples, and people he communicated with, for the most part, spoke Aramaic, the common language of Judea in the first century AD.

The Aramaic predominance and primacy. There is an old theory and assumption that the New Testament was in the beginning written in Greek. New evidence suggests that it was witten in Aramaic, the primary starting place and source: Jesus. The Aramaic New Testament according to George Lamsa was written before the Greek version. Aramic to English translation, Lamsa translation remains the best known of Aramaic to English translation of the New Testament.

Make a note of the present translation (Aramaic to Greek, to Latin, and to English) "My God, my God, why have you forsaken me?" (Mk 15:34) is a quotation in Aramaic of the opening of Psalm chapter 22. It is not what Jesus on the cross, in Aramaic, said and meant. Keep in mind, Christ had Cosmic Consciousness! He spoke, sermonized, and preached in Aramaic and called out, prior to coming back to life (resurrection) in three days, in Aramaic. The Lamsa translation is direct Aramaic to English.

Dearest reader, go back to the original. Go back to Jesus and His Aramaic statement. Go to the infused contemplation and experience yourself what the "motion picture" is all about! Like the laws of physics, the "movie" is unchanging. It is waiting for you to experience it. Christ knew that He came to save the lost. He understood that He was in God and that God was Him. Would Christ proclaim to the world that His Father has forsaken Him? This is fake translation nonsense! Would His apostles be delivering that type of hogwash sales talk and die for it? Notice that Saint John says nothing about it.

To justify the invasion of Poland, which precipitated World War II in Europe, Adolf Hitler set in motion (fake news and killings) Operation Himmler, a series of special SS operations undertaking (i.e., staging false attacks on themselves using concentration camp prisoners and innocent people, affidavit by Alfred Naujocks at the Nuremberg Trials, ethnic cleansing of Germans living in Poland, the Gleiwitz incident, etc.) to create the appearance of Polish aggression against Germany. Jesus understood that "I am the Alpha and the Omega," who is, who was, and who is to come (Rv 1:8)—that is, "Whoever has seen me has seen the Father" (Jn 14:9). In union with God (appendix 1), you realize, as was mentioned earlier, only God exists: "Raise up a stone, and you will find me there." "Father, into your hands I place my spirit"—When He said this, He breathed his last this life cycle for the next three days (Lk 23:46). "It is finished!" (Jn 19:30). He has departed to the total perfection from which He came.

As we have indicated earlier, during His life, Jesus spoke openly about what would happen to Him: crucifixion, death, and then the resurrection. Matthew (16:21–28), Mark (8:31–32), and Luke (9:22–27) cover it. Additionally, to the Jewish authorities, Jesus stated, "Tear down this *house of God* [my emphasis, the temple Jesus spoke of was His body (Jn 2:21)] and in three days I will build it again" (Jn 2:19, Mk 14:58, and Mt 26:61). With Pilate, the chief priest and the Pharisees stated, "Sir, we have recalled that that impostor while he was still alive made the claim, he said, 'After three days I will rise'" (Mt 27:63).

More than seven hundred years before Jesus was born, Isaiah (740–686 BC) the prophet, the author of the longest book of prophecy in the Old Testament, accessed in infused contemplation (predictions) Christ's *early life* ("To whom has the arm of the Lord been revealed, he grew up like a sapling before him, like a shoot from the parched earth" [Is 53:1–2]), *trial* ("He was silent and opened not his mouth, oppressed and condemned, he was taken away" [Is 53:7–8]), *death* ("He surrendered himself to death and was counted among the wicked; and he shall take away the sins of many, and win pardon for their offenses" [Is 53:12]); and the *Resurrection* ("He shall see the light in fullness of days," [Is 53:11]). The Resurrection has been covered by Matthew (28:1–10), Mark (16:1–8), Luke (24:1–12), and John (20:1–10).

"It happened that one of the Twelve [disciples], Thomas [the name means "twin"], was absent when Jesus came. The other disciples kept telling him, 'We have seen the Lord!'"

His answer was, "I will never believe it without probing the nailprints in his hands, without putting my fingers in the nailmarks and my hand into his side." A week later, the disciples were once more in the room, and this time Thomas was with them. Despite the locked doors, Jesus came and stood before them. "Peace be with you," he said; then to Thomas; "Take your finger and examine my hands. Put your hand into my side. Do not persist in your unbelief but believe!" Thomas said in response, "My Lord and my God!" Jesus then said to him: "You become a believer because you saw me. Blest are they who have not seen and have believed." (Jn 20:24–29)

Jesus of Nazareth, being conscious that "I am the Light that is above them all, that the all came from Me," and knowing that going to Jerusalem involved arrest, trial, suffering, crucifixion, passing away, and resurrection (Mt 16:21 and 27:63, Mk 8:31, Lk 9:22, Jn 2:19) must have considered other trouble-free alternatives in His infused contemplation options? There was Cana, Capernaum, Chorazin, Decapolis, Gennesaret, Jericho, Nain, Nazareth, Sidon, Sychar, and Tyre, without facing the establishment—the the chief priests with the elders, the teachers of the Law, and the Sanhedrin. To demonstrate the Resurrection, Jesus required the loss of life. Also, He required maximum broadcasting/dissemination results of His resurrection. He did not care about his pain and death. He knew he will resurrect in three days (see parallel translations for Matthew 18:11):

"For the Son of Man came to save the lost" (American Bible Society);

"For the Son of man came to save the lost" (The Complete Jewish Bible);

"For the Son of man has come to save what was lost" (George Lamsa Translation of the Peshitta); and

"For the Son of Man came to save [from the penalty of eternal death] that which was lost" (The Amplified Bible).

Why? For the "Father in heaven does not want any of these little ones to be lost" (Mt 18:14). Jerusalem is the maximum impact/diffusion decision. Jerusalem was the best and most excellent opportunity to go and make disciples of all nations (Mt 28:16–20) and bring the prodigal sons back home (Lk 15:11–32).

Reminder! Yes, I am the Absolute and the One, my dearest reader, who is with you always. I am the father of everything; I am the mother of pure illimitable light into which no eye can look. I am the sky, the smile, and the sorrow. I am the little ant and the graceful sparrow. I am the child you love and the Christ we execute. I am the goodness-giving goodness, the law, the beginning, and the end. Everyone and everything is me. And there is no one else but me. I am gazing out of every face, including yours. Indeed, I am who am, existing in silence and being prior to everything.

Remember, before this journey, you knew everything. Only youspoke of yourself to yourself. There was no time or space and no birth or death. There was nothing further to be experienced. Timeless eternity and ceaseless bliss seemed boring. You decided to create an illusion to limit the spectrum of your perception. Now you appear to yourself disjoined as a flower, as a waterfall, as a bright sky, or as your own brother, sister, father, mother, or child. Yet you made certain that the pressure of self-discovery takes you back to I AM total knowledge. Do not be uneasy, my love, about this great pilgrimage.

Do not be troubled with so many different faces. They are all carved out of the same one light of the Absolute. They are you!

Rejoice; you will be enraptured in your blissful glory. Yes, my light, you will perceive the triumphant Absolute in your own being.

Know yourself. Press onward to deeper awareness and knowledge of truth. Achieve the limitless freedom. It is a luminous road of radiant beauty, splendor, and light. It is a captivating pilgrimage you have prepared for yourself. It is light. It is love. It is liberty. It is Christ. It is you.

Why Reading by the Candle when You Can Have the Sun

In Salzburg, the city of classic film version of the Rodgers and Hammerstein musical *The Sound of Music*, in addition to conventional bombs, there were delay-action bombs. The delay-action bombs would detonate at different times, with the bomb-time fuzzes set to delay the explosion for times ranging from very brief to weeks, when people were not in the mountain bunkers. Longer delays were intended to disrupt salvage and spread terror in areas where there could still be live bombs and to attack bomb disposal workers.

I vividly remember the Communist (1939–1941) and the Nazi (1941–1944) occupation, gruesome massacres, and atrocities against resident population. The Communist targeted particularly Christian clergy and leaders in Ukraine, which shook me to the core. Cardinal Sosyf Slipyi (1892-1984), who numerous times went through illnesess as well as having his legs and hands broken and frostbitten was imprisoned for 18 years among criminals, investigatos, and jailers in Siberia, Mordovia, and Krasnoyarsk region, and who our children welcomed in the USA, used to characterize the Communist, "with mountains of corpses and rivers of blood".

Pope John Paul II beatified and canonized more saints than every one of his predecessors had throughout the preceding 400 years. Beatification (Latin beatus, "blessed" and "facere" to make) is an acknowledgment accorded by the Catholic Church to have attained the blessedness of heaven and have the capacity to intercede on behalf of individuals who pray in his or her name in Heaven. Canonization officially recognizes the beatified person a saint.

On October 10, 1982, Pope John Paul II canonized his brother, Pole Maximilian Kolbe, from Auschwitz Concentration Camp, which we have referred to earlier. Nineteen years later, on June 27, 2001, Pope John Paul II beatified and canonized twenty-eight Ukrainian martyrs, nuns, priests, bishops, and laymen who died for Christ under the Communist rulers' tyranny. Some saints who had the conscious experience of the Absolute Splendor or understood their divinity are listed in appendix 3.

Following the Soviet's Battle of Stalingrad against the German forces in World War II was one of the largest and bloodiest battles in the history of war (Feb 1943). As the Soviet side started heading west from Stalingrad (now Volgograd), our family did also, leaving all possessions behind except our hearts and deeds, "the workman after all, is worth his keep" (Mt 10:10). We proceeded (1943–1949) from Ukraine via Poland, Czechoslovakia (now Czech Republic and Slovakia), Hungary, Austria, Germany, and to the beautiful America: "God shed his grace on thee, and crown thy good with brotherhood from sea to shining sea."[6]

After arriving to the United States, there were new challenges and new spiritual opportunities: not knowing the English language, no skills or experience for a reasonable job, no money for higher education, severe competition to get into institutions of higher learning, Korean War (June 1950–July 1953) including A-1 draft classification, etc. With those conditions, at age sixteen, how would you feel entering America the Beautiful with spacious skies? I love you! I am profoundly grateful to live in the land of the free.

From my parents and the war experiences, I discovered that all challenges have a maturing, sacred purpose. To all adversities, there is a hallowed solution and an opportunity to grow, develop, and benefit humanity, each of us looking to others' interests rather than his/her own. There is the pure providence of God. Here, I learned to *access* and *experience* the pure providence of God and the Abraham Lincoln's providence—referenced earlier—as a miracle after miracle solution and also to *listen* to the Eternal for the assistance on which I relied, depended in life, and knew God knew what to do in my personal life *beyond my thinking*. This is a state to go *beyond miracles*. This is a life of God on all dimensions. Be at peace of God, which passes all understanding. *Let God be God in you*. All things work together for the good of those who are called according to his decree (purpose) (Rom 8:28). Connect with people on that level.

At college, IBM (International Business Machines) engaged me into the Thomas J. Watson Research Laboratory in Poughkeepsie, New York, when the company was shifting from tabulating machines in Endicott, New York, to transistorized computers in Poughkeepsie.

After testing and evaluating new employees, I was assigned to a four-person team to plan, design, and build a giant Stretch supercomputer (about $100 million in today's currency each) for the Atomic Energy Commission (AEC) and the National Security Agency (NSA).

As the Stretch planning, design, and customer delivery marched to completion, my responsibilities increased. There were travels and family relocations. Here, at the age of twenty-nine, I was IBM's technical director at the NASA Jet Propulsion Laboratory (JPL), California Institute of Technology (Caltech), responsible for the Space Flight Operation Facility development (hardware and software) that controlled *the first soft landing on the moon.*

My wife (who was taken as a five-year-old child to a "slow death" liquidation gulag in Kazakhstan by Stalin forces and later as an eight-year-old by Hitler forces to Dachau Concentration Camp) and I had three children and a mother-in-law, who also survived the Dachau Holocaust encounter. Her husband and brother died in Auschwitz Concentration Camp, three months after their detention.

At first, I meditated now and then, ten to thirty minutes per day. At that time, I did not appreciate, as I do now, the essence or the nature of the Holiest of Holies, eternal, unchanging Alpha and the Omega God Almighty, who is, who was, and who is to come. I believed in God. I did not realize that God is in my shoes, within me, and outside of me. Similar to your hands, legs, head, heart, etc., God is one being. Also, I did not understand how the omnipotent majesty of God, us, and matter integrated into the immense picture of nature.

Now I understand that everything is *the same essence* in different form and amount (zero point field) of Bindu the point: (Jesus's the Kingdom of God as a mustard seed [Mt 13:31–32]) and *nada*, described elsewhere.[7] In addition, I did not value the cosmic power of *focusing* on stillness—that is, restraining mental fluctuations, thoughts, and thinking and also focusing the mind on the divine *peace* that passes all understanding. The stillness synchronizes (harmonizes or attunes) you to the unchanging God, *activating kundalini* to go to the top of the head.

Kundalini, Isaiah's oil of gladness, brings about *expansion of consciousness* in all directions in quietude by bringing into line our consciousness *to* the eternal Absolute. Here, I learned that the highest, uppermost learning you get is your own—the one you learn in stillness alone or, as the most sacred Hindu *Bhagavad-Gita* put it,

> Those who aspire to the state of self-discipline should seek the Self [Godhead] in inner solitude through meditation, controlling body and mind, free from expectations and attachment to material possessions.[8]

In addition to irregular meditations to improve my physical condition, I embarked on a water fast *one day* (twenty-four hours) a week plus once *three days* a month for two years. After two years of this practice, I discovered that my flashbacks vanished, and I was able to walk without arthritic swelling and pain in my knees. Also, following two years of this routine, I experienced more vigor, focus, greater attention span, and clarity of thinking, as well as new energy, the practice of virtue, charity, and detachment. Here, I learned that *we grow to be* what we think or do, and what *we grow to be* we shall attract. If we can change the pattern of thought or action about any situation, then that situation is going to change.

American statesman and self-educated attorney Abraham Lincoln—who guided the country through its most significant constitutional, political, and moral crisis in the gruesome American Civil War and who preserved the Union, abolished the slavery, restructured the US currency and economy, and strengthened the federal government—put it this way: "Most folks are as happy as they make up their minds to be."[9]

William James (1842–1910) was an American philosopher (leading thinker of the late nineteenth century), psychologist (the Father of American psychology), and one of the most influential philosophers of the United States. He observed this:

> The greatest revolution in our generation is the discovery that human beings, by changing the inner attitudes of their minds, can change the outer aspects of their lives.[10]

During college, after four to six hours of working with numbers, accounting or physical equations, I was tired and not very fruitful. At this point at work, I became aware that after twelve hours of intensive effort, I was not tired. My mind was *lucid* and at peace, just as I started. In fact, at that time, *I did not understand what being tired meant.* Here, I realized that my kundalini, the divine energy, was set in motion within my human body.

On the subject of no arthritic swelling and pain in my knees, I shared my experience with a Soviet (Ukrainian) Army general who studied Nazi concentration camp prisoners. He said that concentration camp captives, when freed, were also healed of many ailments before the concentration camp incarceration, with time-restricted feeding and lack of food. Fasting appears to stabilize blood sugar levels, suppress inflammation, decrease blood pressure, improve resting heart rate, and increase resistance to nervous tension and worry. Additionally, fasting can help treat cancer. Cancer will not consume fats. It utilizes glucose as food source. Also, I have noticed fasting improves brain fitness and recall. As a consequence, I have incorporated fasting into my daily life and eating habit. This is comparable to exercise program to get internally in shape.

Sanctity and Higher Degree of Perfection
Consider this an Invitation to Yourself

While my challenges and everyday assignments at work became larger-than-life, I struggled to get out of survival mentality—harnessing de Chardin's narrative of love, purity of intent, simplicity, and dedication and focusing on integrity of character and compassion in my karmic relationship with people. Here is how Pierre Teilhard put it:

> Some day, after mastering the winds, the waves, the tides,
> and gravity, we shall harness for God the energies of love,
> and then, for the second time in the history of the world,
> man will have discovered fire.[1]

Pierre Teilhard de Chardin (1881–1955) was a French philosopher and Jesuit priest who conceived the idea of the Omega Point, incorporating Vladimir Vernadsky's inspiration of *The Biosphere*, an utmost degree of complexity and consciousness toward which the universe is progressing. Notice that as the universe is advancing, there is a fundamental unity of all existence. We are one family in awareness, time, and space. We are responsible for our actions, thoughts, and deeds.

Karma (–: "harmful to you" effects) and *dharma* (+: "beneficial to you" effects) are ones *accumulated* actions, thoughts and deeds in this and previous circumstances of existence. These accumulated actions should be viewed as deciding one's destiny, which is equilibrium. Within the laws of physics and the physical quantities, *equilibrium* is characterized as this: '**1**' (see appendix 2). In the fundamental physical constants form, '**1**' is described as a logarithmic *zero* (0).

John Archibald Wheeler (1911–2008)—whom I met through theoretical physicist Sidney Fernbach (1968), the division chief for the computation division at Lawrence Livermore Laboratory using high-performance Supercomputers to research nuclear weapons, who did a review of *One* for my publisher in 1977, and who studied (appendix 2) paper and the Logarithmic Slide Rule of Physical Relationship—in his beautiful book *At Home in the Universe* writes this:

> The boundary of a boundary is *zero* (my highlighting). This central principle of algebraic topology, identity, triviality, tautology, though it is, is also the unifying theme of Maxwell electrodynamics, Einstein geometrodynamics, and almost every version of modern field theory. That one can get so much from so little, almost everything from almost nothing, inspires hope that we will someday *complete the mathematization of physics and derive everything from nothing, all law from no law* (my emphasis).[2]

John Wheeler is extremely perceptive. '**1**' (the final law of nature, the Absolute, the logarithmic zero: 0) constitutes not only the boundary of a boundary, but is the *cause of all*. There is no other cause. Enlightenment means seeing the Absolute in all things (I AM WHO I AM) and all things in the Absolute, including every aspect of your life. Everyting is derived from *it*: the naught before the Big Bang and Nature, the religious Spirit where we live and move and have our being. Additionally, the zero is independent of the law of causation, thus "thou art That." The *it*, the equilibrium ("=") is all law, yet it is no law.

The final decision is *ours*: the Golden Rule, which we will present shortly applies. Also, when we become more conscious we will see God in all things and all things in God. Also, that we are the creators (performers and audience), as stated earlier, and expanded later:

---

1. Niels Bohr: "We are ourselves both actors and spectators;"
2. John Wheeler: "The universe does not exist 'out there' independent of us;"
3. Arthur Eddington: "We have succeeded in reconstructing the creature that made the footprint. It is our own";
4. Max Planck: "I regard consciousness as fundamental."

"On the other hand," states American theoretical physicist Steven Weinberg, 1979 Nobel Prize for Physics and 1991 the National Medal of Science at the White House,

> Wheeler once remarked that when we come to the final laws of nature, we will wonder why they were not obvious from the beginning. I suspect that Wheeler may be correct, but only because by then we will have been trained by centuries of scientific failures and successes to find these laws obvious.[3]

> If there were anything we could discover in nature that *would* give us some special insight into the handiwork of God, it would have to be the final laws of nature. Knowing these laws, we would have in our possession the book of rules that govern stars and stones and everything else. So it is natural that Stephen Hawking should refer to the laws of nature as "the mind of God."[4]

We harvest what we plant. It's essentially cause and effect the final law of nature. A series of lifetimes where everything is cosmically equilibrated ('**1**') and counts. As Job puts it, "Those who plow for mischief and sow trouble, reap the same. By the breath of God they perish, and by the blast of his wrath they are consumed" (Jb 4:8).

The world's distant cultures, religions, and civilizations at different times of history pointed to the same truth and its idea of the One, '**1**' (manifestations of being), from which all existence emanates and is equilibrated. They expressed the accumulated cosmic past actions and equilibration in as the *Golden Rule* principle of treating others as you would want to be treated (Love all; do wrong to none):

Do not do to others whatever is injurious to yourself. (Zoroastrianism, Zoroaster, Shayast-na-Shayast 13.29)

Treat all creatures in the world as you would like to be treated. (Jainism, Mahavira, Sutrakritanga 1.11.33)

This is the sum of duty: Do not do to others that if done to you would cause you suffering. (Hinduism, Mahabharata 5.1517)

What is hateful to you, do not do to another. That is the entire Teaching; all the rest is commentary. (Judaism, Hillel, Talmud, Shabbat 31a)

Do nothing to others that you yourself would find hurtful. (Buddhism, Buddha, Udanavarga 5.1)

Do not do to others what you do not want done to yourself. (Confucianism, Confucius, Analects 15:23)

Regard your neighbor's gain as your gain, and your neighbor's loss as your loss. (Taoism, Tai-Shang Kan-Yin P'ien)

Do not judge, and you will not be judged. Do not condemn, and you will not be condemned. Pardon [forgive], and you will be pardoned [forgiven]. Give, and it shall be given to you . . . For the measure you measure will be measured back to you. (Christianity, Jesus, Lk 6:37–38)

No one of you is a believer until you desire for others that which you desire for yourself. (Islam, Muhammad, Hadith)

Lay not on any soul a load that you would not wish to be laid upon you, and desire not for anyone the things you would not desire for yourself. (Baha'i, Baha'u'llah)

In appendix 2, you can see how a *Cosmolog* (logarithmic slide rule of physical relationships, or LSPR), instead of accumulated cosmic actions, does forward prediction with physical quantities and the laws of physics.

After two years of irregular ten to thirty minutes per day meditation, I made an effort to meditate and contemplate thirty minutes to two hours each day. I did this on the plane, train, taxi, subway, at night, or very early in the morning. My goal was to learn to be in stillness of the mind and inner peace, *in all places*, utilizing conscious union with the Absolute Being, Absolute Love, and Absolute Void, fully awakened engagement in the activity of everyday life. Here, I noticed that body and breathing can be used to calm the mind and be in union with the Eternal Light.

When I had to solve a complicated challenge, which was at least once a month, I embarked on a one- to four-day water fast. Fasting would bring new insights. It is an excellent means of spiritual advancement. Whilst working, fasting helped to combine (a) *problem definition* (in stillness, I would ask the Higher Self what I need to solve), (b) *careful marshaling of facts*, (c) *infused contemplation*, and (d) *certainty in my heart that there is a solution to the challenge*. With that method, more often than not, I was able to solve the majority of challenges. At this juncture of life, I noticed the *solution* and the source of goodness and miracles is *within oneself*, not introduced from outside.

As I experienced more stillness and inner peace, I also experienced more *depth of insight*, coherence, *clarity*, and *lucidity of thinking*. Clarity and lucidity of thinking became more frequent and ordinary with practice of human kindness, sincerity, and love. Note that in direct experience of the Holy of Holies, infused contemplation seeks to advance the silence of stillness and develop the dormant inner *depths* of the Absolute Splendor. In stillness of the mind (union with the Beloved or Nirvana), you tutor *your* mind to pass beyond mental images and are silent. In infused contemplation, you advance further than the stillness and the Nirvana *to listen* to the still, small voice of God within you. *Take note*, the greater the stillness and silence of the mind, the higher the purity of your consciousness and your inner peace, the more kundalini (bread of life, concealed manna) reaches the brain for network connections and luminosity of your being, and the better the quality of your solution.

Some "easier said than done" problems like the search for the fundamental laws of nature (Steven Weinberg's *Dreams of a Final Theory*, otherwise known as the theory of everything) or the mathematical characterization of the Absolute Splendor (appendix 2), on which I worked for several years after leaving IBM, required more intensive consciousness refining and self-discipline by way of *kundalini purification*, with extensive stillness of the mind, infused contemplation, lengthy fasting, and more focused daily living. Essentially, my *Christ value had to multiply to merit the unification of physics treasure.* Here, after thirty minutes to two hours each day of stillness, I made an undertaking to meditate, deliberate, and contemplate four hours each day. The first building block of time was between 3:00 and 5:00 a.m. The second point in time was before my first meal, which would be 2:00 and 3:00 p.m. And the third block of time would be after nine in the evening.

Note that in Eastern philosophy and the religions of Buddhism, Taoism, Zen and Hinduism, *kundalini* means "snake" or "the serpent power." Jesus characterizes this serpent power, this "tree of life in the middle of the garden and the tree of the knowledge of good and bad" (Gn 2:9), as hidden manna, bread of life, and the fruit of the tree of life. In the Revelation, Jesus states several times "to the one who wins the victory" (Rv 2:26):

> I will see to it that the *victor* [my highlighting] eats from the tree of life which grows in the garden of God. [Also,] to the *victor* [my emphasis], I will give the hidden manna. (Rv 2:7, 17)

This serpent power (the Holy Spirit in Christian terms; 1 Cor 12:1–11), this kundalini, sleeping inner giant in every human being, lies coiled at the base of the spine dormant, in (Sanskrit) *muladhara* chakra. It can be awakened, in the course of Christ-like purified living (Jesus's victor, also appendix 1), by Christ or a saint directly or in a dream. In Matthew, we catch a glimpse of Jesus's awakening power through John the Baptist (incarnate Elijah; Mt 11:14, 17:10–13), who was greater (holier) than any man who has ever lived (Mt 11:11) and also who helped people confess their sins and baptize them in the Jordan River of Judaea:

I baptize you in water for the sake of reform [to show that
you have repented]; but the one who will follow me [Jesus] is
more powerful than I. I am not even fit to carry his sandals.
He it is who will baptize [awaken] you in the Holy Spirit
and fire [kundalini]. His winnowing-fan is in his hand.
(Mt 3:11)

Furthermore, kundalini (the Baptism of the Holy Spirit) can be
awakened and developed in the course of austerities (Moses, Isaiah,
Buddha, Pythagoras, Jesus, Teresa of Avila, John of the Cross, Catherine
of Siena, Francis of Assisi, Gandhi, Mother Teresa, etc., or by way of
transmission of spiritual power [Shakti; dynamic aspect of Godhead]).
Additionally, kundalini can also be stimulated via one who has attained
oneness with God as well as has the power of spiritual awakening and
God-realization. Once awakened, kundalini rises by way of a series of
centers of subtle energy (chakras), referred to as lotuses, through which
the kundalini rises via the spinal channel of *sushumna* to the top of the
head. Kundalini can easily penetrate your blood-brain barrier, where
the thousand-petaled lotus in the brain (Sanskrit *sahasrara*) is located
at the crown of the head. Here, it is pouring out *luminous* white light,
in the form of awareness (Samadhi—a state of meditative union with
the Absolute, Nirvana, or enlightenment), wisdom, knowledge, mystical
visions, power to heal, miracles, and ecstasies of love and bliss (1 Cor
12:4–11).

Note: the blood-brain barrier (BBB) is a wall that surrounds the
brain and prevents harmful toxins, bacteria, and certain important
antioxidants from your vitamins or supplements from reaching your
brain. As a life-saving protection, it also prevents many drugs from
getting to the brain, creating a problem in treating tumors. In the sacred
Holiest of Holies, infused contemplation, we enter from mother or infant
milk to solid food and from luminous light, orgasmic stillness, and
emptiness to the Absolute Void and direct realization. Different planes
of emergence take place from you. They are your own magnificent
grandeur. They are you. Thou art that, I am that, I am indeed the
Absolute Being.

Christianity Has Been Revealed
It Has Not Yet Been Understood or Lived

For individuals interested in the infused contemplation mode used in the state-of-the-art design of the central part of Big Blue Supercomputer (IBM 7030/7950) Stretch/Harvest intended for the Atomic Energy Commission (AEC) at the Los Alamos Scientific Laboratory (LASL) (the top-secret laboratory established in 1942 to design the atom bomb by Robert Oppenheiner, and his team, often referred to as the "father of the atomic bomb").

In addition to LASL, another Big Blue Stretch went to the Lawrence Livermore National Laboratory (LLNL) in California, where the theoretical physicist Sidney Fernbach was and who wanted me to meet co-founder of LLNL and its director, the Hungarian scientists known as the Martians, Edward Teller (1908-2003), also branded informally as "the father of the hydrogen bomb" (where I passed on this opportunity). Some of the Martians: Edward Teller, Eugene Wigner, John von Neumann, Leo Szilard, Theodor Karman, George Polya, Paul Erdos, etc., were leading Hungarian physicists and mathematicians who emigrated from Hungary to the United States in the 20th century.

At this juncture of *infused contemplation*, similar to Wolfgang Mozart, Ludwig Beethoven or Johannn Sebastian Bach pieces of music the following is a snippet from a "Carry Select Adder," the Stretch central CPU, composition:

Hosanna! Here it was. Unexpectedly, like cream rising to the top of milk, the answer surfaced in my mind. An "inner light" of insightful quality illuminated me. I was able to "visually observe" outside of my head a binary tree of *Boolean equations* of thousands of exotic parallel microprocessors. I looked at the internal makeup of parallel super-computers able to add new ten thousand problems to the old ten thousand problems, simultaneously, in parallel. I picked up my pencil and began transcribing the vast string of Boolean relationships [IRE Trans. Electron. Computers (June 1962), patented by IBM in 1963] *that I vividly saw in front of me suspended in the air* above my head.

Note that *all* Boolean equations appeared at once or at the same instant. There were no corrections necessary to the equations. It took me several weeks to comprehend the solution. My manager understood it immediately.

My IBM experiences were remarkable and significant to me as I was exposed to outstanding talent and *penetrating intellect*: IBM had unparalleled, excellent, hardworking, honest, and caring people. After twelve years with the company, I decided to shift from the corporate mode of livelihood in search of higher knowledge, transcendent wisdom, and insight within my own divine light to become a channel of divine grace for others. Should I run out of money, I left the door open with IBM.

Over the years, I proceeded consciously to expand the *ida, pingala,* and *sushumna* channel (the small passage rising from the base of the spine). The Sanskrit *muladhara,* literally "root and basis of existence" through the spine to the midpoint of head—the *sahasrara,* or crown chakra—is the land of the Absolute consciousness. And with growth *outside* the head and body to upper dimensional chakras (higher energies and more developed Christ Nature) in which we live and function, increasing our *quality* and the *quantity* of kundalini spiritual force.

As pointed out earlier, after experiencing constructive results of small-time fasting, I have incorporated fasting into my daily living. Over the years, I frequently fasted from four days to seven days, eleven days, fifteen days, twenty-one days, to as long as forty days with only water. It enabled me to attain deeper inner peace, increase kundalini purity, production, and flow, improve brain well-being and memory, improve high-quality work, compassion, purity of heart, detachment, stillness of the mind, and infused contemplation. On the negative side, I learned that fasting doesn't give you free license to eat or drink whatever you want. As you purify your body, fast foods, meats, and sugar in different products, etc.—like consuming sand—cause pain in your body.

Note: Inner peace, high-quality work, compassion, purity of heart, detachment, stillness of the mind, and infused contemplation will not only *change* your physical appearance and shift you to a higher level of perception but *also* your mind and your natural environment. Your restrictions and limitations of time and space will rupture. You discover that earth's "lower school" inhabitants who reside in our planetary hospital, or dungeon, don't even know they carry the key of the ultimate principle kingdom in their hands.

When the Sun Is Shining, Headlights Are of No Use, Become Divine Light. Fast

What is fasting (see appendix 1, page 123)? Fasting is not only the food for saviors, Doctors of the Church, saints, and prophets, but for all of us in development of the soul, considered the summit of mystical literature. Fasting encompasses food, *habits, emotions, speech, thinking, dealings, events,* and *actions.* "It is the invitation to a secret feast," says Mevlana Jalaluddin Rumi:

> The place where neither I nor space exists . . .
> You are the divine light of the sky.
> If you were to go faster on the way to God,
> You would rise to the sky.
> The throne is your place.
> Aren't you ashamed of dragging yourself
> Like a shadow on the ground?
>
> You will be purified from bad habits by fasting.
> You will follow the attained ascent to the sky by fasting
> You will be burned like a candle by the fire of fasting.
> Become divine light.[1]

Here, employing fasting *with* infused contemplation, I was able to solve and resolve problems of local and global scale and observe into the inner structure of reality and notice *how we, as the creator of our personal universe* (see Krishnamurti in this book, page 85), *project manifestation* via our good and bad thoughts and actions, their effect on the environment and our daily experiences (page 121). In time, it was feasible for me to decode the *sequence* and the *spectrum* of physical quantities and the *fundamental physical constants* with higher precision than can be currently measured. Also, I constructed a logarithmic slide rule for physical relationships (LSPR: Cosmolog) that knows physics (appendix 2). It or a computer can directly generate *verified* laws of physics.

Most significantly, it was possible to see how the unchanging, unmoving Absolute Splendor (our ultimate goal, the Absolute Essence, That Which Is, the Being), is the foundation of nature. To see how the laws of physics and the fundamental physical constants (a connected string of points, Jesus's mustard seed; Mt 13:31, Mk 4:30, and Lk 13:18) are produced. How does the Absolute Light of Certainty ('**1**') enters into *every* law of physics and how the Absolute is in each law of physics *and* action? You can observe this on the LSPR (appendix 2) or as the golden ratio (phi = 1.61803), the constant *e* (2.71828) for compounding financial interest and mathematical modeling of basic life processes as growth or decay, and pi (3.14159) examples in *Celebrate Your Divinity* pages 485–7.

In essence, to succeed in constant union with the Absolute Splendor, one has to live the Absolute Splendor in tranquil body with a liberated mind, craving nothing, mindful and detached, and calm and unperturbed. In fundamental terms, one cannot untie this knot (progress toward God, progress in God, and progress beyond God—permanent Nonbeing) just by listening to stories. One has to be in this world every moment, Christ-like. Jesus says it plainly, "You must be perfect—just as your Father in heaven is perfect. Let your light so shine before men that they may see it, and glorify your Father which is in heaven" (Mt 5:48; Mt 5:16).

Jesus Said, "Let Him Who Seeks, Not Cease Seeking until He Finds, and When He Finds, He Will Be Troubled, and When He Has Been Troubled, He Will Marvel, and He Will Reign over All"

The Gospel according to Thomas is also known as the Coptic Gospel of Thomas. It is a noncanonical collection of 114 sayings of Jesus, who was a profound and fundamental thinker and a daring and heroic campaigner for the Absolute Splendor, the Kingdom of God within us.

Indeed, he shed his own blood for that cause to save the lost. The gospel was discovered near Nag Hammadi, Egypt, in 1945 among a collection of books known as the Nag Hammadi library. *Gospel* means "good news." The gospel is probably the most important resource for our knowledge of the beginnings of Christianity in Egypt. Oscar Cullmann, Lutheran theologian and advisor to three popes, maintains that it is equivalent in significance to the Dead Sea Scrolls and of even greater implication to students of the first three Gospels and their literary sources. "Present documents do not go beyond the end of the second century, whereas this document or its core probably come from the first century," maintains Otto A. Piper, Professor of New Testament Literature and Exegesis at Princeton Theological Seminary. There are 114 *logia*, which represents enumeration within this collection of sayings of Jesus, which comprise the gospel.

Jesus's sayings can be summed up in the Greek word *metanoia*: "I am the way, and the truth, and the life" (Jn 14:6). The term *metanoia* suggests change of mind, regret, conversion, and transformation of consciousness from self-centeredness to God-centeredness, Cosmic Consciousness, illumination, and "God's own peace, which is beyond all understanding,will stand guard over your hearts and minds, in Christ Jesus" (Phil 4:7). Arriving ("The truth will set you free" [Jn 8:32]) is entering the kingdom of heaven—being born again in the spirit. Saint Paul expresses it in the Epistle to the Romans as the renewing of your mind in Christ. It's the exchange of usual way of thinking for God's way of thinking:

> Do not conform yourselves to this age [the standards of this world], but be transformed by the renewal of your mind, so that you may judge what is God's will, what is good, pleasing and perfect. (Rm 12:2)

Let us contemplate these secret words that the living Jesus, who brought about the realization of the identity of oneself and others, spoke and Didymos Judas Thomas wrote:

Let Him Who Seeks,
Not Cease Seeking Until He Finds

The objective of seeking is to discover the Unchanging Being in all that is changing. When one finds the One, one knows the real self of a human being is the Absolute Splendor, which is beyond space and time and gravity; but it is the One Life, the One Being, and Bliss. That realization awakens your being. Here everything that is, is holy ("I am the Absolute Splendor"). The universe is only an outer shell or manifestation. Additionally, one realizes that anything we know, think about, value, experience, are aware of, empathize, figure out, recognize, or understand occurs to us only within consciousness. *Brihadaranyaka Upanishad* expresses it thusly:

> Lead me from the unreal to the real. From darkness lead me
> to light. From death lead me to immortality.[1]

Most people live their own life. So subject to each person's growth and development, area of interest, field of study, and branch of learning, each person will experience different outcomes in their seeking, finding, and *awakening*, commonly understood as enlightenment. Even though awakening from a deep sleep and lack of knowledge of our divinity are different by its nature, they are similar.

All the same, there are relatively distinctive degrees of this experience. If we compare the course of action to breaking through a wall, then the experience can differ between a small pencil-size hole in the wall and the total obliteration of this wall as in the absolute awakening to supreme reality (enlightenment) of Jesus Christ, Shakyamuni Buddha, Isaiah, Lao Tzu, or Meister Eckhart and all the degrees in between. Even though in both cases the same reality is seen, the variations in precision, simplicity, accuracy, clearness, lucidity, intelligibility, and transparency of *insight* are vast.

At one time, I asked my earthy tour guide, "What is this place [planet Earth]?"
His response was "hospital."
When asked, "What do they [from where he is] call it?"
He said, "Hell."

When questioned, "Do they ever get out of here?"

"Yes," was his reply.

When inquired, "What heals them the fastest?"

"Pain," was his answer.

Other individuals had comparable analogies for the overcoming of the barrier that alienates one person from another. Rabindranath Tagore names it dungeon, Rumi prisoners, Sir Jeans cave, Plato chained prisoners in a cave, Isaiah dungeon in darkness and the land of the shadow of death (Is 42:7, 9:2) and also freedom for the captives and release from darkness for the prisoners (Is 61:1), Saint Paul "slavery" (Rom 8:19), Shantideva "the great ocean of suffering," Hasan Lutfi Shushud "the dungeon of existence," and so on. When we realize that the budding kingdom of God (the noble, pure, and saintly Christ, Buddha, and Messiah nature) is in each individual yet mankind has been building nuclear weapons and killing each other, a hospital for sanity and sanctity is not a bad depiction.

The Apostle Paul conveys our oneness, one life, and one being in another way:

> The body is one and has many members, but all the members, many though they are, are one body; and so it is with Christ. It was in one Spirit that all of us, whether Jew or Greek, slave or free were baptized into one body. All of us have been given to drink of the one Spirit. Now the body is not one member, it is many. If the foot should say, "Because I am not a hand I do not belong to the body," would it then no longer belong to the body? If the ear should say, "Because I am not an eye I do not belong to the body," would it then no longer belong to the body? If the body were all eye, what would happen to our hearing? If it were all ear what would happen to our smelling? As it is, God has set each member of the body in the place he wanted it to be. If all the members were alike, where would the body be? . . . God has so constructed the body as to give greater honor to the lowly members, that there may be no dissension in the body, but that all the members may be concerned for one another. If one member suffers, all the members suffer with it; if one member is honored, all the members share the joy. You then,

are the body of Christ. Every one of you is a member of it.
(1 Cor 12:12–27)

Saint Augustine (354–430) was a Roman African, Manichaean, early Christian theologian, and doctor of the church whose writings (*The City of God*, *De doctrina Christiana*, and *Confessions*) influenced the advancement of Western Christianity and also advised the inward journey into the self rather than the outward movement to the world:

> Lord, I have sought you in all the temples of the world and lo, I found you within myself. If a man does not find the Lord within himself, he will surely not find Him in the world.[2]

Saint Catherine of Genoa (1447–1510) was an Italian Catholic saint and mystic. She was also one of the most insightful gazers into the sought after secrets of the Divine Splendor. She continues:

> My "me" is God nor do I recognize any other "me" except my God.[3]

From Jesus's life, we know that He was a Revealer of Truth. He came to earth to make known, disclose and reveal our forgotten identity, our Divinity. Now you can share the heavenly nature and God's divine power in you. At the very commencement in the public life in the Gospels according to John, Matthew, and Luke, Jesus made this known:

> The Father and I are one. (Jn 10:30)
> The Father is in me and I am in him. (Jn 10:38)
> All that the Father has belongs to me. (Jn 16:16)
> Whoever looks on me is seeing him who sent me. (Jn 12:45)
> Is it not written in your law, I have said, "You are gods"? If it calls those men gods to whom God's word was addressed—and Scripture cannot lose its force. (Jn 10:34–35)
> I am in my Father, and you are in me, and *I in you* [my accent]. (Jn 14:20)

Combining John 10:38, 12:45, and 14:20, it denotes that whoever looks on you is seeing the Father, the Unchanging, the Alpha and the

Omega God Almighty, who is, who was, and who is to come, as the manifest, the Absolute Splendor, in human flesh of you!

Albert Einstein (1879–1955), the German-born theoretical physicist who had a breakthrough penetrating the mind, revealed the law of the photoelectric effect, a basic solution in the advance of quantum theory, the mass-energy equivalence relationship, the Bose-Einstein statistics, etc.; and he spent the later part of his life searching for a single theoretical framework to express the oneness of all and the unification of physics (i.e., the unified field theory) and described a transient, temporary, *impermanent* universe of time and space and gravity with his general theory of relativity and the special theory of relativity. In Einstein's *oneness*, to this time, he had not understood *himself* to be the Absolute Splendor or life.

That is, in Einstein's unification papers, he appeared to understand the oneness of the material, manifest Universe; however, the *unmanifest* Spirit, the unmanifest Absolute Splendor, and therefore us have not been integrated into the unified field theory, the fundamental law of nature, the Steven Weinberg's dream of a final theory equation: '**1**' ("For in *him* we live and move and have our *Being*" [my highlighting; Acts 17:28]). And Christ (the image of the invisible God) is the visible likeness of the invisible God (Col 1:13-15).

Notice that Einstein, with his unified theory, and Saint Paul both address impermanence. However, following a deeper, more profound understanding of our oneness, Saint Paul makes his public statement in the Epistle to the Galatians:

> The life I live now is not my own; Christ is living in me."
> (Gal 2:20)

Furthermore, Saint Augustine, Saint Catherine of Genoa, and Jesus give us examples of those who found the Absolute Splendor within themselves.

The reign of God on earth is like the Kentucky treasure hidden in the field of your being. That being, like the oil field, has to be developed by upgrading and improving your heart, mind, and life. The development of one's heart and mind overcome all physical barriers to bring all individuals and all things into a harmonious communion.

Then we will realize it is no longer "I" that live, but it is the eternal, imperishable, unmanifest Most High as manifest the Absolute Splendor who lives as us. After one finds the Absolute Splendor, she or he helps others reach the same goal.

And When He Finds, He Will Be Troubled

What does *troubled* mean? What does anxious, concerned, bothered, worried, disturbed, distressed, uneasy, unsettled, and unfortunate mean? Similar to breaking through a wall, from a small pencil puncture or a total obliteration of the wall example, there are gradations of and perception of a much deeper dimension within us being troubled.

On one level of *awakening*, you start seeing things you did not see before. For example, you perceive that life in this dimension is essentially one of suffering—which the Buddha, Jesus, Muhammad and others have acknowledged. You recognize Jesus's statement: "As often as you did something for one of my least brothers, you did it for me" (Mt 25:40). You become conscious that all your actions are karmically equilibrated and affect not merely an earthly human but God who is within that human. That may as well trouble you.

What's more, you awaken to the impermanence of nature, which by the way is also a central concept of Buddhism—that is, you apprehend that all composite things (time, space, motion, protons, neutrons, atoms, energy, matter, relativity) are the void, empty, temporary, devoid of an essence (Sanskrit *shunyata*, Japanese *ku*), as all creation arises from the mind of the Absolute Splendor: '1' (Godhead, That Which Was Never Born, That Which Never Dies, and absolutely immutable Being). That which is first and foremost reality apart from whose existence nothing else could be or be conceived. That may be another type of concern.

Moreover, you become conscious that *emptiness does not mean nonexistence*. It is *not* nothingness. For instance, a computer may be reset to zero. Its registers may be empty—void of manifestation. Because of their conditioned character, they are created and therefore considered to be the *cosmic void*, the Absolute Splendor, from which all things emerge and to which they return, such that the essence of all things is the Absolute Splendor, a nowness of emptiness, which you yourself are—as the entire universe is emptiness (appendix 1).

You also awaken to the ultimate fate of the cosmos! In the Absolute Splendor awareness, you clearly see that you live in a *personal universe* and the lengths of your path (an encoded hologram: a multidimensional field of points) *depend on you* and that after billions of years or incarnations, the universe will end in a big crunch, as the galaxies and clusters would slowly break down into a particle soup inverse to that in the big bang, and the universe returns to the *singularity*, the Omega (Rv 1:8) (also Pierre Teilhard de Chardin's Omega Point), from which it was born. This cycle (the Absolute Splendor adventure, exploration, and traveling around) has happened infinite times before, and it will happen infinite times again.

In the Absolute Splendor awareness, a person *awakens* to the eternalness, immovability, and *permanence* of the Absolute—which is beyond all categories of time, space, and causality. The Gospel of John depicts,

> Before the world was created, the Word [Christ] already existed, he was with God, and he was the same as God. From the very beginning, the Word was with God. Through him God made all things; not one thing in all creation was made without him. The Word was the source of life, and this life brought light to men. (Jn 1:1–5 abs)

The past, the now, and the future is here in a very, very tiny *empty* point. It was one of the toughest questions I tried to penetrate. It was hard to grasp how the whole universe and its history with other universes can be in a point of void. You realize that the power of our mind can be multiplied by many planes of infinity through technology, spiritual capacities, and Cosmic Consciousness. You also understand that the creative source of life, when the world was created, has been transported through the equilibrium state (=) to the miracle zone and into the manifest world.

Notice that although now we are able to see galaxies billions of light years away, we are also at the "cave" age in attaining union with our Being of beings (the ultimate principle of all reality) rather than through rituals, sacrifices, wars, Soviet and Nazi concentration camps, the nuclear bomb arsenal, etc. At this juncture, we understand that the Kingdom of God (the place of life) is within you, outside you, and everybody as well.

There is much we have to say about this matter, but it is hard to explain and to understand (Heb 5:11). As revealed earlier, everything that is, is holy, sacred, and divine. However, you may contemplate whether the planet is ready for such a profound advancement and global transcendence beyond religious conviction, beyond *infants milk* to *solid food* for adults and the mature (Heb 5:11–14). You may also find it a challenge to share it with others. When my book *One*, which discusses our oneness, was published in 1977, it had only my first name, Orest.

In the Gospel according to Thomas, the 80 and 81 *logia*, which represent the enumeration within this collection of sayings of Jesus, we read,

> The kingdom is within you, and it is outside of you. When you know yourselves, then you will be known, and you will know that you are the sons of the living Father. But if you do not know yourselves, then you are in poverty, and you are poverty.[1]

On December 6, 1273, Saint Thomas Aquinas was celebrating a morning Mass in *infused contemplation*. Following the event, Saint Aquinas was clearly disturbed and stopped writing his *Summa*. When asked to carry on, Saint Aquinas stated,

> I can do no more; such things have been revealed to me that all I have written seems as straw, and I now await the end of my life.[2]

It is a challenge to get a mature elephant to journey through a two-inch opening or an eighteen- wheeler truck through a ten-inch crevice. How many times did Saint Aquinas read these statements and not appreciate their significance, usefulness, and consequence?

"To you the mystery [secret] of the reign [kingdom] of God has been confided [is given to you]. To the others outside it is all presented in parables" (Mk 4:11).

"I am in my Father, and you are in me, and I [am] in you" (Jn 14:20). "The spirit of the Lord fills the world [whole universe], is all-embracing, [and holds all things together], and knows what man says" (Wis 1:7). "They know not, neither do they understand; they go about in darkness; all the foundations of the earth are shaken. I said: You are gods; all of you are sons of the Most High; yet like men you shall die, and fall like any prince. Rise, O God, judge the earth, for yours are all the nations" (Ps 82:5–8).

"God the Logos became what we are, in order that we may become what He Himself is" (Saint Irenaeus).

"You are the light of the world" (Mt 5:14).

On that same day, December 6, 1273, Saint Thomas Aquinas, in infused contemplation, spiritually awoke from a deep sleep. In purity of intent and stillness of the mind (like widening the openings for the elephant and the eighteen-wheeler truck), he understood the I AM: "I am who I am" (Ex 3:14), I AM that, I AM indeed the Absolute Being, I AM perfect in my own being. Saint Thomas Aquinas understood why, by means of purity of heart and stillness of the mind, Jesus was doing miracle after miracle: enlightening the blind, raising the dead, walking upon the waters of the sea dry-shod, rebuking the winds, cleaning lepers, healing the palsied, and many other profound wonders. Like Saint John of the Cross, Aquinas now could shout from the rooftops:

> Mine are the heavens and mine is the earth. Mine are the nations, the just are mine, and mine are the sinners. The angels are mine, and the Mother of God, and all things are mine; and God Himself is mine and for me, because Christ is mine and all for me. What do you ask for? What do you ask, then, and seek, my soul? Yours is all of this, and all is for you.[3]

Or Mechthild of Magdeburg:

> The day of my spiritual awakening was the day I saw and knew I saw all things in God and God in all things.[4]

> I am in you and you are in me. We cannot be closer. We are two united, poured into a single form by an eternal fusion.[5]

As written, *Summa* makes no common sense; it misses the key point. It does not tell the reader in the Spirit of the solid food: "I am in you, and you are in me. We cannot be closer." It does not enlighten "men who have the Spirit" (1 Cor 3:2; Heb 5:12–14), the future priests, bishops, cardinals, and so forth:

> Brothers, the trouble was that I could not talk to you as spiritual men but only as men of flesh, as infants in Christ. I fed you with milk, and not solid food because you were not ready for it. You are not ready for it even now, being still very much in a natural condition [of this world]. (1 Cor 3:1–3)

> Solid food is for the mature, for those whose faculties are trained by practice to distinguish good from evil. (Heb 5:11–14)

> Let us, then go beyond the initial teachings about Christ and advance to maturity, not laying the foundation all over again [and leave behind us the first lessons of the Christian message]. (Heb 6:1)

Now Saint Aquinas's statement "I can do no more" has meaning. "Such things [that] have been revealed to me that all I have written seems as straw" is precise, correct, and truthful. "And I now await the end of my life." Three months later, when Saint Aquinas was forty-nine and passed away, it should not have taken place. His superiors should have understood that God is *within* each of us, Emmanuel ("God with us," God-among-us, and the oneness of all creation) escorts us past the existing intellectual awareness that divides us, one from another, and segregates us into hostile, fighting clans. Having arrived at the eternal world, with neither beginning nor end, one becomes conscious that we are all one and that there is no me without you or you without me because you are me (God in human flesh and appearance). In infused contemplation this is what happened to Saint Aquinas.

Since the *Summa* is used to develop and guide future spiritual educators (professors, priests, bishops, archbishops, cardinals, and popes), the whole *Summa* should have been written in infused contemplation. This is what the scientific world does. Without knowing, it strives to verify with measurements *the eternal world* of the unchanging and the laws of physics. Saint Aquinas could have rewritten the *Summa*. The Catholic Church, the planet, and humanity would have advanced, benefited, and uplifted where Jesus, Saint Paul, Moses, Muhammad, the Buddha, Lao Tzu, Nanak, Baha'u'llah, and others (appendix 3) sought us to be—building heaven on earth.

I Am: I Am Who I Am

From Moses, we learn that I AM ("I am who I am" [Ex 3:14]) freed the Israel from the torment of relative reality and the dungeon of existence in Egypt. And afterward we find out from Jesus that the Kingdom of God *is* within you, within each one of us. Thus, there is no idol worshipping on the way of Absolute liberation and Cosmic Consciousness. There is only compassionate thank-you. Leading with purpose, we're all human.

In his 1971 message, His Holiness Pope Paul VI reminded us forcefully of the fundamental truth that "every man is my brother." Then how can we kill (thou shalt not kill, not even in the name of nation), steal, commit adultery, bear false witness against anyone, desire thy neighbor's wife, covet thy neighbor's house, his field, or anything that is thy neighbor's when you know God called himself "I am who I am" (Ex 3:14)?

After forty days and forty nights without eating or drinking, he had no sandals, for the place where he was standing was holy ground—on Horeb, the Mountain of God. Moses comes down the mountain with *the two stone tablets of the covenant*. The tablets were the work of God, and the writing was the writing of God, engraved on the tablets ("I AM: I AM THAT I AM"). God says, "This is my name forever; this is my title for all generations" (Ex 3:14).

As Moses approached the camp and saw the sculptured, idol calf and the celebratory dancing, he was in agony and disturbed. In reverence for life, he knew his people were not ready *the way God pointed out to them*—for "I AM: I AM THAT I AM" reasoning, living, and the two tablets of the covenant. They were not enlightened to Cosmic Consciousness. They needed to mature to function as Christ on earth.

Following the Egypt experiences, Moses's heart must have experienced ache and distress. Keep in mind, subsequent to forty days and forty nights without eating or *drinking*, Moses's body must have been in pain and suffering. He threw the *"I AM: I AM THAT I AM"* writing of God in tablets out of his hands, smashing them at the base of the mountain. Then he took the calf statue they had made, burned it in the fire, ground it to powder, and scattered the powder over the face of the water. Then he forced the Israelites to drink it (Ex 32:20).

Jesus's "He will be troubled" is on the mark! Nonetheless, there is a shortcut: to reach peace, teach peace and the Ten Commandments. On the other hand, *awakening* to "I AM: I AM THAT I AM" and *living* at "I AM: I AM THAT I AM" are the keys to escape.

And When He Has Been Troubled, He Will Marvel

Where is the marvel? Where is the wonder? In short, you will marvel when knowingly you experience the Bhagavad-Gita proverb "In thyself know thy friend, in thyself know thy enemy." Team up for excellence; you treat all beings as you would treat yourself. Furthermore, it means you will marvel when consciously you see yourself, others, and the whole universe as the Absolute Splendor or as the *Upanishads* said,

Tat tvam asi: That art Thou: Thou art That.[1]

Or "this entire world is Brahman ('**1**')." "Yes, a new form of humanity most beautiful and most wonderful is being evolved right under our eyes," states the philosopher of the United Nations and its prophet of hope, in *New Genesis*, Robert Muller. Saint Paul described that state as "the life I live now is not my own; Christ is living in me" (Gal 2:20).

And He Will Reign Over All

In Matthew we are made aware by Jesus: "Reform your lives! The kingdom of heaven is at hand" (Mt 4:17). People who have transformed their lives give us a beautiful overview of their direct encounter and union with God (the life in which our will is united with God).

You will rule over all when you realize that your true self is the Absolute Splendor, which is beyond space-time, and that your highest goal is self-realization: union with God—the final triumph of the Spirit, the flower of humanity's crown note. In union with God, you leave yourself in the arms of your Father and *let His will be your will*. Let His Law be your law. Let "not I, but Christ in me" be your motto (appendix 1, page 126). In this manner, you have engaged *consciously* the Absolute to be your essence, your mind, your law, your action. You are its eyes, its ears, its hands, and its legs.

Thus, you have entered into the unmanifest and imperishable mode of pure consciousness, associated with a feeling of beyond-words joy, as well as the scholarly understanding. You are liberated; you are free. At this point, your human life has developed into living your Absolute Splendor—a perfect manifestation of divine essence.

To know truth is to become truth, for the essence of the self is *conscious divinity*—the light of God. As the Leviticus and Saint Paul in the Second Epistle to the Corinthians stated:

> I will set my Dwelling among you, and will not disdain you. Ever present in your midst, I will be your God, and you will be my people; for it is I, the LORD, your God, who brought you out of the land of the Egyptians and freed you from the slavery, breaking the yoke they had laid upon you and letting you walk erect. (Lv 26:11–13)

> You are the temple of the living God, just as God has said: "I will dwell with them and walk among them. I will be their God, and they shall be my people." (2 Cor 6:16)

Saint Athanasius the Great (AD 296–373), Archbishop of Alexandria, in Cosmic Consciousness (a higher form of consciousness than that possessed by the ordinary person [simple consciousness or self-consciousness]), realized that "God became man so that we might become children of God." This means that the Absolute Splendor became one of us so that we can become what God is: divine.

It should be recognized that not all cases of Cosmic Consciousness are equivalent. Some are more prominent and more perfect. The more number of years one is subjected in Cosmic Consciousness, the higher the quality and quantity and more perfect the results.

Saint Gregory of Nazianzus (AD 329–390) called Athanasius the pillar of the Church. Athanasius is counted as one of the four great Eastern Doctors of the Church. In the Eastern Orthodox Church, he is labeled as the Father of Orthodoxy.

Some Signs of the Cosmic Sense

Fear of death vanishes.
One knows without learning.
Instantaneous intellectual illumination.
New life: spiritual rebirth and moral elevation.
The sense of timelessness, eternal life, and immortality.
Access to other tracks, other lives, and other reincarnations.
Consciously becoming one being with That Which Is: God.
Increased powers of accumulating knowledge and initiating action.
Added faculty: smell, color, music creativity, perception, and consciousness.
You enter a deep, deep ocean of peace and joy within, which passes all human understanding.

Richard Maurice Bucke (1837–1902) was a contemporary of William James (1842–1910). At age thirty-five, the Canadian doctor experienced cosmic sense (consciousness). In Christian mysticism, Buddhism, and Brahmanism, it is a prize of extensive and rigorous self-discipline and endeavor. The account of the experience is quoted from the *Proceedings and Transactions of the Royal Society of Canada*:

> He and two friends had spent the evening reading Wordsworth, Shelley, Keats, Browning, and especially Whitman. They parted at midnight, and he had a long drive in a hansom. His mind, deeply under the influence of the ideas, images and emotions called up by the reading and talk of the evening, was calm and peaceful. He was in a state of quiet, almost passive, enjoyment.
>
> All at once, without warning of any kind, he found himself wrapped around, as it were, by a flame-colored cloud. For

an instant he thought of fire—some sudden conflagration in the great city. The next [instant] he knew that the light was within himself. Directly after there came upon him a sense of exultation, of immense joyousness, accompanied or immediately followed by an intellectual illumination quite impossible to describe. Into his brain streamed one momentary lightning-flash of the Brahmic Bliss, leaving thence forward for always an after-taste of Heaven.[57]

In 1901, nearly thirty years after his illumination, Bucke published the book *Cosmic Consciousness*.[58] This classic study in the evolution of the human mind is a very important addition in the saga of human growth and development. Professor William James read *Cosmic Consciousness* shortly following its appearance and wrote to the author:

I believe that you have brought this kind of consciousness "home" to the attention of students of human nature in a way so definite and unescapable that it will be impossible hence forward to overlook it or ignore it . . . But my total reaction on your book, my dear Sir, is that it is an addition to psychology of first rate importance, and that you are a benefactor of us all.[59]

Based on the evidence of three thousand years of human history, Bucke shows fourteen instances of Cosmic Consciousness and that, in addition to them, there have been other instances of partial or debatable Cosmic Consciousness, noticing the rising frequency of the experience. Bucke concluded that the human race is in the progression of developing a new kind of consciousness, far in advance of the common self-consciousness.

Instances of Cosmic Consciousness Identified by Bucke:

Gautama the Buddha
Jesus the Christ
Saint Paul
Plotinus
Mohammed
Dante

Bartolomé de Las Casas
John Yepes
Francis
Jacob Behmen
William Blake
Honore de Balzac
Walt Whitman
Edward Carpenter

The Greatest Achievement
Cosmic Consciousness

The greatest achievement[1] of all time and the ultimate miracle on earth is not our governments, our laws, our institutions, or our agricultural mechanization. It is not our sciences, religions, commerce, finance, water supply, or distribution. It is not our supercomputing, thermonuclear weapons, jets, television, genetic engineering, or our powerful secrets to achieving superior financial returns and high income. It is not our medicine, communication, transportation, material wealth, space stations, landing on the moon, or our discovery of the secrets of the stars.

Like no other treasure on earth, the greatest achievement of all time and the ultimate miracle on earth is the *conscious experience of our Absolute Splendor*, '**1**'. Yes, the wonder of wonders: the Cosmic Consciousness validation and confirmation of the fundamental unity and holiness and purity of all existence.

Here you bring the past into the present. Here you bring the future into the present. All is now! Every part of the eternity, as the Vedic *Bhagavad-Gita* and Saint Symeon the New Theologian put it, with *infused contemplation* is accessible to every single one now.

Is that right? Yes, the Nirvana (*moksha, beatific vision*) encounter of the *absolute equilibrium*, the most fundamental reality, the ultimate underpinning for the infinite-storied tower of the final law of nature, the "Jerusalem above," and the I AM ("He who exists . . . he who is") are in us as us!

Additionally, it is the *miracle zone*, '**1**', the easy way, and *the fundamental consciousness* that abides in all. It is the first cause that opens infinite doorways for the human spirit and for the integration of humankind into the multiverse society to a profound pathway of unlimited freedom and peace and a most abundant life of joy, happiness, and unlimited opportunities.

In appendix 3, you will find reproducible results, the proof and forward guidance of the greatest achievement of all time and the ultimate miracle on earth. It contains a list of 406^2 remarkable individuals, exalted liberators, giants of the Allah-Buddha-Christ-Krishna spirit, including many scientists from all over the world who personally verified the Most High, "I AM," and the fundamental consciousness (Cosmic Consciousness) in themselves.

They recognized the holiness of life—that we are many colors of one sacred rainbow. They recognized that we are *one single Being*—the "I AM HE WHO IS" (Ex 3:14). Most of these individuals in different countries and poles apart of history, in most cases without knowing about one another's stunning evidence, have independently described their results and conclusions. They have verified and corroborated our collective unity within the nature and development of our spiritual consciousness in unchanging character of the Absolute Splendor (the reality). In the "Summation of Appendix 3," you can see the 406-list distributions by countries.

Summation of Appendix 3
Verification, Corroboration, and Evidence by Countries

India 73	Tibet 32
China 31	Ukraine 30
Korea 29	USA 29
Egypt 15	England 15
Germany 15	France 12
Iran 12	Japan 12
Italy 11	Austria 6
Greece 6	Turkey 6
Israel 5	Nepal 5
Pakistan 5	Russia 5
Spain 5	Afghanistan 2
Belgium 2	Burma 2
Iraq 2	Latvia 2
Lebanon 2	Poland 2
Sweden 2	Albania 1
Algeria 1	Arabia 1
Armenia 1	Australia 1
Bangladesh 1	Bulgaria 1
Canada 1	Croatia 1
Czech Rep. 1	Denmark 1
Hungary 1	Ireland 1
Jamaica 1	Kashmir 1
Nederland 1	New Zealand 1
Slovakia 1	Sri Lanka 1
Switzerland 1	Syria 1
Tajikistan 1	Thailand 1
Vietnam 1	Uzbekistan 1[3]

Waking the Illumination of the Self and Miracles

Who am I? Why am I here? Where am I going? Can I cultivate the hidden potentiality of my being? What is life all about? Is there some kind of enhancement of awareness or expansion of perceptual and experiential horizons to see things as they are? Buddha, Jesus, Muhammad, Nanak, Baha'u'llah, and many others who have personally built a knowledge base about our Absolute Splendor have provided paths to the spiritual awakening—which is a way for beauty, wisdom, harmony, peace, prosperity, health and bliss. Most of these precious inheritance is in the open, yet it is not understood by most of us. Reality is unchanging, but *our perception of reality changes as we change our awareness. We see what we see, and we know what we know because we are who we are.* To access our most valued inheritance, we have to invest into our own being to seek God, to see God, and to experience God. The great Apostle Paul elaborates on what mindset will distinguish the new person:

> If anyone is in Christ, he is a new creation. The old order [creation] has passed away; now all is new! (2 Cor 5.17)

In 1911, Evelyn Underhill wrote a without-comparison, all-purpose presentation to the lessons of mysticism: *Mysticism* (the mystic fact and the mystic way). This remarkable book is the classis work in its field and is an unsurpassed general introduction to the study of the nature and development of our spiritual consciousness. It addresses the soul's ascent (progressive apprehension) of reality (Godhead, '**1**'):

The awakening of the self.
The purification of the self.
Voices and visions.
Recollection and quiet.
Contemplation.
Ecstasy and rapture.
The dark night of the soul.
The unitive life (absolute immersion in God).

Approximately one hundred years later (2008), Professor Luke Timothy Johnson, former Benedictine monk and teacher at Yale Divinity School, wrote thirty-six lectures on *Mystical Tradition: Judaism, Christianity and Islam*:

A Way into the Mystic Ways of the West.
Family Resemblances and Differences.
The Biblical Roots of Western Mysticism.
Mysticism in Early Judaism.
Merkabah Mysticism.
The Hasidism of Medieval Germany.
The Beginnings of Kabbalah.
Mature Kabbalah—Zohar.
Isaac Luria and Safed Spirituality.

Sabbatai Zevi and Messianic Mysticism.
The Ba'al Shem Tov and the New Hasidism.
Mysticism in Contemporary Judaism.
Mystical Elements in the New Testament.
Gnostic Christianity.
The Spirituality of the Desert.
Shaping Christian Mysticism in the East.
Eastern Monks and the Hesychastic Tradition.
The Mysticism of Western Monasticism.
Medieval Female Mystics.

Mendicants as Mystics.
English Mystics of the 14th Century.
15th- and 16th- Century Spanish Mystics.
Mysticism among Protestant Reformers.
Mystical Expressions in Protestantism.
20th Century Mystics.
Muhammad the Prophet as Mystic.
The House of Islam.

The Mystical Sect—Shi'a.
The Appearance of Sufism.
Early Sufi Masters.
The Limits of Mysticism—Al-Ghazzali.
Two Masters, Two Streams.
Sufism in 12th–14th Century North Africa.
Sufi Saints of Persia and India.
The Continuing Sufi Tradition.
Mysticism in the West Today.

Professor Johnson has done marvelous "pure intellectual stimulation" in *Mystical Tradition: Judaism, Christianity and Islam* "that can be popped into the [audio or video player] any time," states *Harvard Magazine*. Evelyn Underhill maps out her view of the mystic's journey into "Awakening of Self," "Purgation of Self," "Illumination," "the Dark Night of the Soul," and the unitative life.

Though Underhill is focused on mysticism in Christianity, she also brings up in *Mysticism* Buddhism, Hinduism, Sufism, and other traditions. In a breathtaking episode, distinctive in the writing of mysticism, Angela of Foligno has described the eloquent revelation in which she perceived this truth—the revelation of the Absolute in deep humility and unspeakable power realizing ibn Arabi's insight that "there is no other existence than He," that "all the world is Brahman" as Upanishads relates:

> The eyes of my soul were opened, and I beheld the plenitude of God, wherein I did comprehend the whole world, both here and beyond the sea, and the abyss and ocean all things. In all these things I beheld naught save the divine power, in a manner assuredly indescribable; so that through excess of marveling the soul cried with a loud voice, saying, "This whole world is full of God!"[1]

Yes, "this whole world is full of God," states Angela of Foligno. As the direct experience and the scientific evidence confirm, nothing can exist outside of God; therefore, everything is already united to God ('**1**'), as the laws of physics, and the high precion fundamental physical constants prove.

Where Do We Have To Go?
Blessed Are the Pure in Heart,
for They Shall See God

Many Lives and Countless Times

Many lives have you had as insects and worms.
And many lives as elephants, fish and deer;
In many lives were you born a snake or a bird,
And countless times, you lived as a tree.
After aeons you've obtained the glory of human birth:
Now it's your chance—meet the Lord![1]

—Guru Arjan
The Fifth Guru of Sikhism

I Am Awake

A man came across the Buddha on the road one day and was awestruck.
"Are you a god?" he asked.
"No" was the reply.
"Are you an immortal?"
"No."
"A holy saint?"
Once again, the Buddha answered, "No."
"Then," the man asked, "what are you?"
The Buddha simply said, "I am awake."[2]

—Traditional Buddhist Story

Arthur Eddington (1882–1944) was a foremost English astrophysicist, mathematician, philosopher of science, and one of the world's greatest astronomers. He achieved fame as the very first person to investigate the internal structure and the natural limit to the *luminosity* of the radiation generated by accretion onto a compact object (named in his honor), evolution, and motion of the stars.

Eddington was also the first expositor of the general theory of relativity (GTR), the geometric theory of gravitation published by Albert Einstein 1915, and the bending of light as a result of gravity. Like with the bending of light due to gravity and luminosity of light due to stars, Eddington concluded:

> We have found that where science has progressed the farthest the mind has regained from nature that which the mind has put into nature. We have found a strange footprint on the shores of the unknown. We have devised profound theories, one after another, to account for its origin. At last, we have succeeded in reconstructing the creature that made the footprint. And lo! It is our own.[3]

David Joseph Bohm (1917–1992) was a very important American (from a Hungarian Jewish father and a Lithuanian Jewish mother) theoretical physicist of the twentieth century. He taught physics at Princeton University, University of Sao Paulo, University of Bristol, and the University of London's Birkbeck College.

Bohm associated with Jiddu Krishnamurti, Robert Oppenheimer, Richard Feynman, Isidor Rabi, David Pines, Yakir Aharonov, etc. He contributed to the nature of reality, quantum physics, neuropsychology, the nature of consciousness, the functioning of thought, and oneness.

Bohm understood that the *empty space* is the foundation and the ground for the existence of everything including ourselves:

> The entire universe has to be understood as a single undivided whole, in which analysis into separately and independently existent parts has no fundamental status. . . . What we perceive through the senses as empty space is actually the plenum, which is the ground for the existence of everything, including ourselves. The things that appear to our senses are derivative forms, and their true meaning can be seen only when we consider the plenum, in which they are generated and sustained, and into which they must ultimately vanish.[4]

> What I am proposing here is that man's general way of thinking of the totality, his general worldview, is crucial for overall order of the human mind itself. If he thinks of the totality as constituted of independent fragments, then that is how his mind will tend to operate, but if he can include everything coherently and harmoniously in an overall whole that is undivided, unbroken, and without a border [for every border is a division or break] then his mind will tend to move in a similar way, and from this will flow an orderly action within the whole.[5]

Max Karl Ernst Ludwig Planck (1858–1947) was an exceedingly important and prudent German theoretical physicist. He is the originator of quantum theory, which revolutionized our understanding of atomic and subatomic structure of nature.

Planck was known for his constant (the Planck constant) and the Planck postulate (the electromagnetic energy is produced only in invariable [Planck constant] packets), for which he received the Nobel Prize in Physics.

He is also recognized for his Planck's law of black body radiation, the Fokker-Planck equation, the Nernst-Planck equation, and the third law of thermodynamics. In addition to the Nobel Prize award in Physics (1918), Planck received Foreign Associate of the National Academy of Sciences (1926), Lorentz Medal (1927), Copley Medal (1929), Max Planck Medal (1929), and Goethe Prize (1945).

In 1948, the Kaiser Wilhelm Society, the German scientific society of which Planck was twice president, was renamed to Max Planck Society (MPS). The MPS is one of the four big science organizations within Germany.

The MPS currently consists of eighty-three institutes in and outside of Germany, representing a wide range of scientific objective, which carry out basic research in life science, natural sciences, and social and human sciences. Looking at nature, Planck stated,

> I regard consciousness as fundamental. I regard matter as derivative from consciousness. We cannot get behind consciousness. Everything that we talk about, everything that we regard as existing postulates consciousness.[6]

John Archibald Wheeler was a renowned American theoretical physicist. He worked on unified field theory, general relativity, and geodynamics; help design and build the hydrogen bomb; and supervised forty-six PhDs (more than any other professor in the Princeton physics department).

Wheeler received numerous awards: Einstein Prize (2003), Wolf Prize (1997), Matteucci Medal (1993), Albert Einstein Medal (1988), J. Robert Oppenheimer Memorial Prize (1984), Oersted Medal (1983), National Medal of Science (1970), Franklin Medal (1969), Enrico Fermi Award (1968), and Albert Einstein Award (1965).

Similar to Arthur Eddington creature that made the footprint, Wheeler concluded that we live in a participatory universe where observations of the universe (the anthropic principle) must be congruent with the conscious and harmonizing life that perceives it:

> The universe does not exist "out there" independent of us. We are inescapably involved in bringing about that which appears to be happening. We are not only observers. We are participators. In some strange sense this is a participatory universe. Physics is no longer satisfied with insights only into particles, fields of force, into geometry, or even into time and space. Today we demand of physics some understanding of existence itself.[7]

> We are participators in bringing into being not only the near and here but the far away and long ago. We are in this sense, participators in bringing about something of the universe in distant past, and if we have one explanation for what's happening in the distant past why should we need more?[8]

Niels Henrik Bohr (1885–1962) was a very commanding Danish physicist who made fundamental contributions to understanding atomic structure and quantum theory, for which he received the Nobel Prize in Physics in1922. Similar to the Buddha and David Bohm, Bohr realized that everything we describe as existent or real is made of things that cannot be regarded as existent or real. Using Lord Jesus's lingo, they are created and therefore temporary, like a wave in the ocean.

In addition to the Nobel Prize, Bohr received the Hughes Medal (1921), the Matteucci Medal (1923), the Franklin Medal (1926), the Copley Medal (1938), the Order of the Elephant (1947), the Atoms for Peace Award (1957), and the Sonning Prize (1961).

Comparable to John Wheeler's participatory universe, Bohr expressed his partaking involvement in the universe by being together with *performers* or artists and *audience* or viewers:

> When searching for harmony in life one must never forget in the drama of existence we are ourselves both actors and spectators.[9]

Erwin Schrödinger (1887–1961) was a foremost Austrian physicist who contributed to different fields and was the author of approximately fifty publications in statistical mechanics and thermodynamics, physics of dielectrics, color theory, cosmology, electrodynamics, general relativity, genetics, ethics, religion, theoretical biology, and the phenomenon of life (*What is Life?*; *What is Life?: With Mind and Matter* [great science classics of the twentieth century]).

Schrödinger received the Nobel Prize (together with Paul Dirac) in Physics (1933) for his formulation of the Schrödinger equation. The Schrödinger equation enables one to calculate the wave function of a system and how it changes dynamically in time. Schrödinger was recognized with the Haitinger Prize (1920) of the Austrian Academy of Sciences, the Matteucci Medal (1927), the Max Planck Medal (1937), the Erwin Schrödinger Prize of the Austrian Academy of Sciences (1956), and the Austrian Decoration for Science and Art (1957).

Schrödinger thought his exact scientific work was an advance to the Godhead. He attempted to integrate various branches of physics (gravity, electromagnetism, and nuclear forces within the fundamental structure of general relativity) into a unity of knowledge, feeling, and choice by means of unified field theory of the eternal and unchangeable Godhead or nothingness:

> What is it that has called you so suddenly out of nothingness to enjoy for a brief while a spectacle which remains quite indifferent to you? The conditions for your existence are almost as old as the rocks . . . Looking and thinking in that manner you may suddenly come to see, in a flash, the profound rightness of the basic conviction in Vedanta: it is not possible that this unity of knowledge, feeling, and choice which you call your own should have sprung into being from nothingness at a given moment not so long ago; rather this knowledge, feeling, and choice are essentially eternal and unchangeable and numerically one in all men, nay in all sensitive beings. But not in this sense—that you are a part, a piece, of an eternal, infinite being, an aspect or modification of it.[10]

Willis W. Harman (1918–1997) was an American electrical engineer, physicist, futurist, and author who associated with the "human potential and life is hallowed" movement. Willis taught at the University of Florida and the Stanford University electrical engineering and physics. In addition, Harmon developed a popular Stanford graduate seminar called The Human Potential that covered topics ranging from meditation to psychoactive drugs to parapsychology. He was made a regent of the University of California by then governor Jerry Brown. Willis served as a regent for ten years.

After leaving Stanford, Harman became a senior social scientist at SRI International and director of SRI's Educational Policy Research center. From 1978 until his death in 1997, he served as president of Institute of Noetic Sciences. For Harman,

> A noetic science—a science of consciousness and the world of inner experience—is the most promising contemporary framework within which to carry on fundamental moral inquiry which stable human societies have always had to place at the center of their concerns . . .[11]

> When reality is wholeness, there is no greater error than separateness thinking. Imagine if the stomach were to get the idea that it could pursue its self-interest independent of the wellbeing of the whole body-mind-spirit. Justifying its appetites with the maxims: "What's good for the stomach is good for the whole," and "The business of the stomach is growth of the stomach," it seeks to maximize its absorption of nutrients and minimize the fraction going to other parts of the body. It worries about such indicators as getting its "market share" of food value, and "gross abdominal product." It sounds absurd, of course, because the stomach doesn't do anything of the sort. It concentrates on performing its function with regard to the whole system, and trusts that if it does that, the system will see to it that its nutrient, protection, and other needs are met.[12]

Rabbi Abraham Joshua Heschel (1907–1972) was a renowned Polish-born American rabbi and one of the foremost Jewish theologians and philosophers of the twentieth century. Heschel arrived in New York City in 1940. He taught at Hebrew Union College in Cincinnati for five years. In 1946, Rabbi Heschel served as a professor of Jewish ethics and mysticism at the Jewish Theological Seminary of America until his death in 1972.

He published Der Shem Hamefoyrosh: Mentsch (1933, a volume of Yiddish poems), The Sabbath (1951), Man is Not Alone (1951), God in Search of Man (1955), The Prophets (1962), Torah min Ha Shamayim (1962), and Prophetic Inspiration after the Prophets (1966).

Like Jesus, the Buddha and others, Heschel's aim was to wake up humanity out of severe ignorance and ultimate embarrassment to our holy and ineffable Absolute Splendor:

> How embarrassing for man to be the greatest miracle on earth and not to understand it! How embarrassing for man to live in the shadow of greatness and ignore it, to be a *contemporary of God* [emphasis mine] and not sense it. Religion depends upon what man does with his ultimate embarrassment.[13]

> Remember that there is meaning beyond absurdity. Know that every deed counts, that every word is power . . . Above all, remember that you must build your life as if it were a work of art.[14]

Jiddu Krishnamurti (1895–1986) was an Indian philosopher, speaker, and writer. His interests included the foundation of nature, the nature of mind, meditation, the liberation of human beings, and lovers of truth.

Krishnamurti wrote The Awakening of Intelligence, Think on These Things, The Flight of the Eagle, Freedom from the Known, and The First and the Last Freedom.

Krishnamurti had association and discussions with physicist David Bohm (who was mentioned earlier), Fritjof Capra, E. C. George Sudarshan, and biologist Rupert Sheldrake, representing an assortment of fundamental interpretations:

> I had the first most extraordinary experience. There was a man mending the road; that man was myself; the pick axe he held was myself; the very stone which he was breaking up was a part of me; the tender blade of grass was my very being, and the tree beside the man was myself. I almost could feel and think like the road maker, and I could feel the wind sighing through the tree, and the little ant on the blade of grass I could feel. The birds, the dust, and the very noise were part of me. Just then there was a car passing by at some distance; I was the driver, the engine, and the tires; as the car went further away from me, I was going away from myself. I was in everything, or rather everything was me, inanimate and animate, the mountain, the worm, and all breathing things.
>
> I was supremely happy, for I had seen. Nothing could ever be the same. *I have drunk at the clear and pure waters at the source of the fountain of life and my thirst was appeased* [my highlighting; see Jesus's water and everlasting life, Jn 4.14, above]. Never more could I be thirsty, never more could I be in utter darkness. I have seen the Light. I have touched compassion which heals all sorrow and suffering; it is not for myself, but for the world.[15]

John White is an internationally known author and editor in the fields of consciousness research and higher human development. He has published fifteen books, including *The Meeting of Science and Spirit*, *What is Enlightenment?*, and *A Practical Guide to Death and Dying*. White is on the boards of various academic and spiritual organizations and is an editorial contributor to national publications. He acknowledged the following:

There is a profound paradox in the spiritual journey. It is this: the goal of our journey, the answer we seek is none other than what we *already are* in essence—Being, the ultimate wholeness that is the source and ground of all Becoming. *Enlightenment is realization of the truth of Being*. Our native condition, our true self is Being, traditionally called God, the Cosmic Person, the Supreme Being, the One-in-all. [Incidentally, some enlightened teachers—Buddha was one—prefer to avoid theistic terms in order to communicate better. Their intent is to bypass the deep cultural conditioning that occurs through such language and blocks understanding]. We are manifestations of Being, but like the cosmos itself, we are also in the process of Becoming—always growing, changing, developing, evolving to higher and higher states that ever more beautifully express the perfection of the source of existence. Thus, we are not only human beings; we are also human be-comings. Enlightenment is understanding the perfect poise of being-amid-becoming . . . When we finally understand that Great Mystery, we discover our true nature, the Supreme Identity, the Self of all. That direct perception of our oneness with the infinite, that noetic realization of our identity with the divine is the source of all happiness, all goodness, all beauty, all truth. The experience is beyond time, space, and causality; it is beyond ego and all socially conditioned sense of "I." Knowing ourselves to be timeless, boundless, and therefore cosmically free ends the illusion of separateness and all the painful, destructive defenses we erect, individually and societally, to preserve the ego-illusion at the expense of others, The *Maitrayana Upanishad* puts it this way: "Having realized his own self as the Self, a person becomes selfless . . . This is the highest mystery."[16]

Jesus said: Blessed are the solitary and elect, for you shall find the Kingdom; because you come from it, [and] you shall go there again.[17] (logia 49, line 26)

also,

Jesus said, "If they say to you: 'From where have you originated?' say to them: *We have come from the Light, where the Light has originated through itself* [emphas is mine].' It [stood] and it revealed itself in their image. If they say to you: '[Who] are you?' say: 'We are His sons and we are the elect of the Living Father.' If they ask you: 'What is the sign of your Father *in you*?' say to them: 'It is a movement and a rest.'"[18] (logia 50, line 30)

and,

Jesus said: I am the Light that is above everything. I am the All, the All came forth from Me and the All has returned to Me. Cleave a [piece of] wood, I am there; lift up the stone and you will find Me there.[19] (logia 77, line 22)

in addition,

The Kingdom of the Father is spread upon the earth and men do not see it.[20] (logia 11)

—The Gospel According to Thomas
(Also known as the Coptic Gospel of Thomas)

You are woman. You are man. You are the son and also the daughter. As an old man you walk with a stick. Being born you assume many faces in all directions. You are the dark blue bird; you are the green parrot with red colored eyes. You are the cloud with lightning in your womb. You are the seasons and the oceans, without a beginning. Because of your powers you move everywhere. All the worlds are born from you.[21]

—The Svetasvatara Upanishad
(An ancient Sanskrit text)

This world is full of God, and everything you see is God.

Everything is held together by God's strength.

God is the intelligence in an insect, the faithfulness in a dog, and the latent energy in a rock.

You are not a man; you are God yourself.

Don't be under the illusion that God is somewhere else and you have to search for him.

I am the seed of all elements and all beings.

The universe is the body of God; every particle in it fills with God, his glory, his might, and his inscrutability.

You need not search for God. You yourself are divine. How can you go in search of yourself? This is the mistake you commit. When everything is permeated by the divine, who is the searcher of the divine? It is because the world has lacked men who could proclaim this Vedantic truth with authentic experience that it has sunk to such degrading levels.

Love All, Serve All. Help Ever, Hurt Never.[22]

—Sri Sathya Sai Baba
(Reincarnation of Sai Baba of Shirdi)

I died to the mineral state and became a plant,
I died to the vegetable state and reached animality.
I died to the animal state and became a man.
Then what should I fear? I have never become less from dying?
At the next charge [forward] I will die to human nature,
So that I may lift [my] head and wings [and soar] among the angels,
but even from angelhood I must pass on
all except God ['**1**'] perishes.
When I have sacrificed my angel soul,

I shall become what no mind e'er conceived.
Then I will become non-existent;
for non-existenvie says to me in organ tones,
Truly, to him we shall return.[23] (*Masnavi* 56)

—Jalaluddin Rumi
(Persian poet, Islamic scholar, Sufi mystic)

The seed of God is in us. Given an intelligent and hardworking farmer, it will thrive and grow up to God, whose seed it is; and accordingly its fruit will be God-nature. Pear seed grow into pear trees, nut seeds into nut trees, and God seeds into God.[24]

—Meister Eckhart
(German theologian and mystic)

I am life which wills to live, and I exist in the midst of life which wills to live. . . . The deeper we look into nature, the more we recognize that it is full of life, and the more profoundly we know that all life is a secret and that we are united with all life that is in nature.[25]

Man can no longer live his life for himself alone. We realize that all life is valuable and that we are united to all this life. From this knowledge comes our spiritual relationship to the universe.[26]

—Albert Schweitzer
(Theologian, philosopher, physician)

Every being is divine, is God. Every soul is sun covered over with clouds of ignorance; the difference between soul and soul is owing to the difference in density of these layers of cloud.[27]

—Swami Vivekananda
(Indian Hindu mystic and saint)

You are a Divine Elephant with amnesia trying to live in an ant hole. Sweetheart, O Sweetheart you are God hiding from yourself.[28]

—Hafiz
(Persian poet and mystic)

Wherever my travels may lead, paradise is where I am.[29]

—Voltaire
(French Enlightenment writer, mystic)

Every man is a divinity in disguise, a god playing the fool.[30]

—Ralph Waldo Emerson
(American essayist, lecturer, philosopher)

There is only one question. And once you know the answer to that question, there are no more to ask. . . . Out of the depths of unbroken Infinity arose the Question, Who am I? And to that Question there is only one Answer—I am God![31]

The problem is that people do not know who they really are: you are Infinite. You are really everywhere; but you think you are the body, and therefore consider yourself limited. If you look within and experience your own soul in its true nature, you will realize that you are Infinite and beyond all creation.[32]

To love God in the most practical way is to love our fellow beings. If we feel for others in the same way as we feel for our own dear ones, we love God.[33]

I speak eternally. The voice that is heard deep within the soul is My voice—the voice of inspiration, of intuition, of guidance. Through those who are receptive to this voice, I speak.[34]

—Meher Baba

(Indian spiritual teacher)

We are not human beings having a spiritual experience. We are spiritual beings having a human experience.[35]

—Pierre Teilhard de Chardin
(French philosopher and Jesuit priest)

Be content with what you have; rejoice in the way things are. When you realize there is nothing lacking, the whole world belongs to you.[36]

—LaoTzu
(Chinese philosopher and writer)

Let nothing upset you. Let nothing frighten you. Everything is changing; God alone is changeless. Patience attains the goal! Who has God lacks nothing; God alone fills every need.[37]

—Saint Teresa of Avila
(Carmelite nun, mystic)

Do not store up for yourselves treasures on earth, where moth and rust destroy, and where thieves break in and steal. But store up for ourselves treasures in heaven, where moth and rust do not destroy, and where thieves do not break in and steal. For where your treasure is, there your heart will be also.[38] (Mt 6:19–21)

—Jesus Christ
(Savior, Jewish enlightened teacher)

It is he who is revealed in every face, sought in every sign, gazed upon by every eye, worshipped in every object of worship, and pursued in the unseen and the visible. Not a single one of His creatures can fail to find Him in its primordial and original nature.[39]

There is no Truth superior to me. Everything rests upon me ['1'], as pearls are strung on a thread. I am the taste of water, the light of the sun and the moon, the syllable om in the Vedic mantras; I am the sound in ether [the empty space—the rest frame] and the ability in man. I am the original fragrance of the earth, and I am the heat in fire. I am the life of all that lives, and I am the penance of all ascetics. I am the original seed of all existence, the intelligence of the intelligent, and the prowess of all-powerful men. All states of being—be they of goodness, passion or ignorance—are manifested by my energy. I am, in one sense, everything—but I am independent. I am not under the modes of this material nature.[40] (7.7–12)

—The Bhagavad Gita
("The Song of God," Sanskrit Scripture)

How Do We Get There?
Be Still and Know That I AM God

In God we trust

Nothing is the same, as fear threatens to divide us;
Yet everything's the same when we all pull together.

Nothing is the same, with the enemy at the door;
Yet everything's the same, for we've all been here before.

Nothing is the same, as nations struggle to survive;
Yet everything's the same in freedom's enduring truth.

Nothing is the same in a world devoured by greed;
Yet everything's the same in the heart of those who love.

Nothing is the same, as men and women die alone;
Yet everything's the same in each soul's eternal home.

Nothing is the same, as arrogance and power loom;
Yet everything's the same as each citizen stands tall.

Nothing is the same, as we seek equilibrium;
Yet everything's the same, for our peace still lies within.

Nothing is the same, as the whole world turns upside down;
Yet everything's the same when we give our lives to God.

—Barbara Benjamin
(Carmelite, mystic, and
Christ Conscious Leadership author)

The quotes in this part of the book are meant expressly for you: to help you enter the great infinity of beauty, wisdom, harmony, and peace and meet the Eternal Light within yourself.

We did what we could. Now it is up to you: the Eternal Light within yourself. In boundless peace, stillness, and solitude, connect with the limitless peace, stillness and solitude within yourself as much and as long and as often as possible.

"Settle yourself in solitude, and you will come upon Him in yourself," advises Carmelite nun Saint Teresa of Avila.

"Seek first the kingdom of God and all else shall be added unto you," said Jesus.

"When will you begin that long journey into yourself?" whispered Rumi.

"Be still and know that I am God." (Ps 46:10)

Of all the hard facts of science, I know of none more solid and fundamental than the fact that if you inhibit thought [and persevere], you come at length to a region of [the principle *above* the laws of nature, the absolute equilibrium, Godhead, and the fundamental] consciousness below or behind thought . . . and a realization of an altogether vaster self than that to which we are accustomed. And since the ordinary consciousness with which we are concerned in ordinary life is before all things founded on the little local self, and is in fact self-consciousness in the ordinary self and the ordinary world. It is to die in the ordinary sense, but in another sense, it is to wake up [from death] and find that one's real, most intimate self, pervades the universe and all other beings. . . . So great, so splendid is this experience, that it may be said that all minor questions and doubts fall away in the face of it; and certain it is that in thousands and thousands of cases the fact of this having come even once to a man has completely revolutionized his subsequent life and outlook on the world.[1]

—Edward Carpenter
(British poet)

When you have closed your doors, and darkened your room, remember never to say that you are alone, for you are not alone; God is within, and your genius is within—and what need have they of light to see what you are doing.[2]

—Epictetus
(Greek philosopher)

If you wish to grow in your spiritual life, you must not allow yourself to be caught up in the workings of the world; you must find alone, away from the noise and confusion, away from the allure of power and wealth.[3]

—Thomas á Kempis
(Augustinian monk)

It is not necessary to off on a tour of great cathedrals in order to find Deity. Look within. You have to sit still to do it.[4]

—Albert Schweitzer
(German physician, missionary, and Nobel Prize winner)

The beating heart of the universe is holy joy.[5]

—Martin Buber
(Austrian philosopher and mystic)

You turn inward. There's nothing to distract you, so you begin to look at yourself.[6]

—Frank Bianco, American
(Journalist/photographer)

The world would be happier if men had the same capacity to be silent that they have to speak.[7]

—Baruch Spinoza
(Dutch philosopher)

In short, every age has need of "contemplative life," and ours is no exception to the rule. The soul needs a chance for spreading its wings, for looking beyond itself, beyond the immediate environment, and for quiet inner growth.[8]

—James Bissett Pratt
(American philosopher)

Keep knocking and the joy inside will eventually open a window and look out to see who is there.[9]

—Jalaluddin Rumi
(Persian poet, Islamic scholar, and Sufi mystic)

Solitude is when you discover God firsthand. You don't need an intermediary.[10]

—Buckminster Fuller
(American architect, philosopher)

Do all the good you can,
By all the means you can,
In all the ways you can,
In all the places you can,
To all the people you can,
As long as ever you can.[11]

—John Wesley
(English cleric, theologian, and evangelist)

When you arise in the morning give thanks for the morning light, for your life and strength. Give thanks for your food and the joy of living.[12]

—Chief Tecumseh
(Native American Shawnee warrior and chief)

If the only prayer you ever say in your entire life is "thank you," it will be enough.[13]

—Meister Eckhart
(German theologian and mystic)

He who offers to me with devotion a leaf, a flower, a fruit, or even a little water, that offering of devotion I accept from him whose self is pure.

Whatever you do, whatever you eat, whatever you offer, whatever you give, whatever austerities you perform, do that as an offering to me.[14]

—The Bhagavad Gita
(Hindu holy book)

When eating a fruit, think of the person who planted the tree.[15]

—Vietnamese Proverb

Believe nothing just because a so-called wise person said it.
Believe nothing just because a belief is generally held.
Believe nothing just because it is said in ancient books.
Believe nothing just because it is said to be of divine origin.
Believe nothing just because someone else believes it.
But believe only just when you yourself test and judge to be true.[16]

—The Gautama Buddha
(Sage of the Shakyas)

If you want others to be happy, practice compassion. If you want to be happy, practice compassion.[17]

—Dalai Lama XIV
(Important monk of the Gelug School)

God does not look at your forms and your possessions, but He looks at your hearts and your deeds.[18]

—Muhammad
(Arab prophet, founder of Islam)

There is nothing as easy as denouncing. It don't take much to see that something is wrong, but it takes some eyesight to see what will put it right again.[19]

—Will Rogers
(American entertainer and humorist)

If you meditate on your ideal, you will acquire its nature. If you think of God day and night, you will acquire the nature of God.[20]

—Sri Ramakrishna Paramahamsa
(Indian Hindu mystic and saint)

Where shall the word be found, where will the word Resound? Not here, there is not enough silence.[21]

—T. S. Eliot
(British American poet)

I prize the privilege of being alone.[22]

—Carl Rogers
(American psychologist)

You need solitude if you are to fulfill your promises.[23]

—Frances Steloff
(Founder of Gotham Book Mart)

We must, like a painter, take time to stand back from our work, to be still, and thus see what's what . . . True repose is standing back to survey the activities that fill our days.[24]

—William McNamara

(American philosopher)

In solitude one can achieve a good relationship with oneself.[25]

—May Sarton
(American poet, novelist, and essayist)

If a woman is to know herself, then periods of solitude should be courted, planned, and embraced.[26]

—Mary Kay Blakely
(American humorist)

Only in the oasis of silence can we drink deeply from our inner cup of wisdom.[27]

—Sue Patton Thoele
(American psychotherapist)

And at some time in your life trying to be good may be to stop running and take time . . . to be quiet and discover who you are and where you've been.[28]

—Sister Corita Kent
(Carmelite nun)

The insight we gain from solitude has very little to do with the amount of time we spend alone. It has a lot more to do with the quality of time we spend with ourselves.[29]

—Jan Johnson Drantell
(American counselor)

Being alone gives us the space to listen again to our inner rhythms, to embrace our inner selves.[30]

—Patricia Hoolihan
(American writer on spiritual concerns)

Solitude is simply spending time connecting with ourselves. Solitude means we do it alone, spending time in reflection—perhaps talking to ourselves, writing a journal, meditating. When we practice solitude regularly over a period of time, we develop a deep and abiding connection with our self. We can use that connection to alleviate isolation—from ourselves and others.[31]

—Jan Johnson Drantell
(American counselor)

Learn to get in touch with silence within yourself and know that everything in this life has purpose. There are no mistakes, no coincidences; all events are blessings given to us to learn from.[32]

—Elizabeth Kübler-Ross
(Swiss American psychiatrist)

The more faithfully you listen to the voice within you, the better you will hear what is sounding outside. Only he who listens can speak.[33]

—Dag Hammarskjöld
(Swedish statesman)

O Solitude, the soul's best friend, that man acquainted with himself does make.[34]

—Charles Cotton
(British poet)

Solitude is the furnace of transformation. Without solitude we remain victims of our society and continue to be entangled in the illusion of the false self.[35]

—Henri J. M. Nouwen
(Professor, writer, and theologian)

You think that I am impoverishing myself by withdrawing from men, but in my solitude I have woven for myself a silken web or chrysalis and, nymph-like, shall ere long burst forth a more perfect creature, fitted for a higher society.[36]

—Henry David Thoreau
(American naturalist)

It is in deep solitude that I find the gentleness with which I can truly love my brothers. The more solitary I am the more affection I have for them. It is pure affection, and filled with reverence for the solitude of others. Solitude and silence teach me to love my brothers for what they are, not for what they say.[37]

—Thomas Merton
(French-born Trappist monk)

Devote six years to your work but in the seventh go into solitude or among strangers so that your friends, by remembering what you were, do not prevent you from being what you have become.[38]

—Leo Szilard
Hungarian physicist)

Silence is precious, for it is of God. In silence all God's acts are done; in silence alone can his voice be heard and his word spoken.[39]

—Mother Mary Madelena

(Carmelite nun)

In the attitude of silence the soul finds the path in a clearer light, and what is elusive and deceptive resolves itself into crystal clearness. Our life is a long and arduous quest after Truth and the soul requires inward restfulness to attain its full height.[40]

—Mahatma Gandhi
(Indian spiritual and political leader)

We must learn to soundproof the heart against the intruding noises of the public world in order to hear what God has to say.[41]

—Gordon MacDonald
(President, InterVarsity Christian Fellowship)

On every mountain height is rest.[42]

—Goethe
(German dramatist)

When we are unable to find tranquility within ourselves, it is useless to seek it elsewhere.[43]

—Rochefoucauld
(French philosopher)

Silence is a friend who will never betray.[44]

—Confucius
(Chinese teacher of wisdom)

Let your mind be quiet, realizing the beauty of the world, and the immense, the boundless treasures that it holds in store. All that you have within you, all that your heart desires, all that your Nature so specially fits you for—that or the counterpart of it waits embedded in the great Whole, for you. It will surely come to you.[45]

—Edward Carpenter
(British poet)

Peace of mind must come in its own time, as the waters settle themselves into clearness; you can no more filter your mind into purity than you can compress it into calmness; you must keep it pure if you will have it pure, and throw no stones into it, if you would have it quiet.[46]

—John Ruskin
(British essayist)

Every kind of creative work demands solitude, and being alone, constructively alone, is a prerequisite for every phase of the creative process.[47]

—Barbara Powell
(American psychologist)

Great ideas come into the world as gently as doves. Perhaps, then, if we listen attentively, we shall hear, amid the uproar of empires and nations, a faint flutter of wings, the gentle stirrings of life and hope.[48]

—Albert Camus
(Algerian French philosophical essayist)

If you are a writer you locate yourself behind a wall of silence and, no matter what you are doing, driving a car or walking or doing housework . . . you can still be writing, because you have that space.[49]

—Joyce Carol Oates
(American novelist)

When the mind is very quiet, completely still, when there is not a movement of thought and therefore no experience, no observer, then that very stillness has its own creative understanding. In that stillness the mind is transformed into something else.[50]

—Jiddu Krishnamurti
(Indian philosopher)

A public man, though he is necessarily available at many times, must learn to hide. If he is always available, he is not worth enough when he is available.[51]

—Elton Trueblood
(American historian)

The hours which I have spent alone with Mr. Edison have brought me the real big returns of my life: to it I attribute all I have accomplished.[52]

—Thomas Edison
(American inventor)

Without great solitude no serious work is possible.[53]

—Pablo Picasso
(Spanish painter)

I prefer to get up very early in the morning and work. I don't want to speak to anybody or see anybody. Perfect silence. I work until the vein is out.[54]

—Katherine Anne Porter
(American novelist)

There is nobody else like you. The more you can quiet your own thoughts, fears, doubts, and suspicious, the more will be revealed to you from the higher realms of imagination, intuition, and inspiration.[55]

—Kenneth Wydro
(American lecturer)

Obviously, if we are to experience insights from our consciousness, we need to be able to give ourselves to solitude.[56]

—Rollo May
(American Existentialist psychologist)

Talent is nurtured in solitude; character is formed on the stormy billows of the world.[57]

—Goethe
(German dramatist)

Solitude is the nurse of enthusiasm, enthusiasm is the true part of genius.[58]

—Isaac D'Israeli
(British essayist)

The best remedy for those who are afraid, lonely, or unhappy is to go outside, somewhere where they can be quite alone with the heavens, nature and God.[59]

—Anne Frank
(German diarist)

It all adds up to one thing: Peace, silence, solitude. The world and its noise are out of sight and far away. Forest and field, sun and wind and sky, earth and water, all speak the same silent language.[60]

—Thomas Merton
(French-born Trappist monk)

When you pray, go into your room and shut the door and pray to your Father who is in secret; and your Father who sees in secret will reward you.[61]

—Jesus Christ
(Savior, Jewish enlightened teacher)

When I need solitude, I turn off the phone and fax and sit until my breath comes slow and gentle, and I am able to enter the sanctuary that always awaits me at the center of my being.[62]

—Sam Keen
(American spiritual teacher)

Within you there is a stillness and a sanctuary to which you can retreat at any time and be yourself.[63]

—Hermann Hesse
(German novelist, poet)

What a strange power there is in silence. How many resolutions are formed, how many sublime conquests effected, during that pause when lips are closed, and the soul secretly feels the eye of her Maker upon her! They are the strong ones who know how to keep silence when it is a pain and grief unto them, and who give time to their own souls to wax strong against temptation.[64]

—Ralph Waldo Emerson
(American pastor, essayist)

The victories of speech have been many, but the victories of silence have been more. The man of silence is the man of power.[65]

—Carl Sandburg
(American poet)

There is something greater and purer than mouth utters. Silence illuminates our souls, whispers to our hearts, and brings them together.[66]

—Kahlil Gibran
(Lebanese American painter, poet)

He that would live in peace and ease must not speak all he knows nor judge all he sees.[67]

—Benjamin Franklin
(American statesman, author)

There are times when silence is the most sacred of responses.[68]

—Eugene Kennedy
(American philosopher)

Better to remain silent and be thought a fool than to speak out and remove all doubt.[69]

—Abraham Lincoln
Sixteenth American President

Keep silent, because the world of silence is a vast fullness.[70]

—Rumi
(Persian poet, Islamic scholar, Sufi mystic)

Even a fool who keeps silent is considered wise.[71]

—Proverbs 17:28

Silence alone can bring two hearts closer together.[72]

—Karl Pruter
(American bishop)

Talking is a loss of power.[73]

—Frederick W. Faber
(British rector)

Silence is unceasing eloquence. It is the best language.[74]

—Ramana Maharshi
(Spiritual teacher, India)

EPILOGUE

An epilogue (after Greek *epilogos*, "conclusion"; from *epi*, "in addition," and *logos*, "word") at this point is the final part of work where the author steps in and "speaks freely and openly" to the reader as afterword.

Exodus 3:14 is the most profound and the most important statement in the entire Old and New Testament Bible. Here God *Himself* acknowledged (made known) to Moses *and* humanity (1400 BC) that "I AM WHO I AM" (or else, "I AM" for short) or the Being of beings, otherwise the Being in each being. The traditional English translation within Judaism favors "I am the Existing One" or "I will be what I will be."

Jesus Christ characterized the "I AM WHO I AM" as "I AM," otherwise "The kingdom of God is within you" (namely, the wave is a part of the ocean) or "whoever has seen me has seen the Father" (simply, whoever has seen the wave has also seen the ocean). Had the elders, the chief priests, and the scribes [of the Law] *experienced* and understood the "I AM WHO I AM" meaning, our planetary history would be different. Jesus would not have been put through the suffering, crucifixion, and death. Had Adolf Hitler, Joseph Stalin, and Mao Zedong experienced and understood the "I AM WHO I AM" meaning and implication to their nations, 150–200 million people would not have perished because of them.

The Upanishads are the oldest Sanskrit texts of spiritual knowledge of Hinduism. They were written between 800 BC and 500 BC. The Upanishads state, "All the world is Brahman" ("I AM," the "Absolute"— in Hinduism, the ultimate reality underlying all phenomena) and "This [whatever exists], too, is Brahman."

The Taoist religion was founded by Lao Tzu who lived 604–531 BC. Taoism was adapted as a state religion in China about 440 BC. Taoism emphasizes living in harmony with the Tao (the "I AM," the '1'). It insists that "there is nothing outside the Tao; you cannot deviate from it."

Islam continues Allah, literally "the God," is the Absolute One, the All-Knowing and All-Powerful. "Everything is perishing save his Face"—that is, there is nothing except his Face—"then, whithersoever you turn, there is the Face of God."

Now by means of the laws of physics and mathematics ('1') we can see that the "I AM WHO I AM" findings of Judaism, Christianity, Hinduism, Taoism, Buddhism, Islam, etc. are correct. Basically God Himself takes on human form and lives as us in the world. That is, before the big bang, singularity (the unmanifest emptiness, '1',) created itself (became a manifest fundamental quantity of time, space, or gravity). As Jesus stated in Thomas: "I am the Light that is above everything. I am the All, the All came forth from Me and the All has returned to Me. Cleave a [piece of] wood, I am there; lift up the stone and you will find Me there" (logia 77, line 22).

The holy books of all religions tell us how to *connect* (fuse, unite, synchronize) with our Being, '1'. Also, earlier we referenced Eddington, Planck, Wheeler, and Bohr, who made basic contributions to comprehension of atomic arrangement and *quantum theory*, for which he received the Nobel Prize in Physics:

1. Arthur Eddington: "We have succeeded in reconstructing the creature that made the footprint. It is our own";
2. Max Planck: "I regard consciousness as fundamental;"
3. John Wheeler: "The universe does not exist 'out there' independent of us;"
4. Niels Bohr: "We are ourselves both actors and spectators."

Yes. "We are ourselves both actors and spectators," and "we see what we see and we know what we know because we are who we are". Individuals who desire to improve and advance their life (the cycle of rebirths stretching back many billions of years without perceptible beginning), liberty, freedom, and pursuit of happiness are suggested to *find more peace within themselves*. Through constant *inner peace* (appendix 1) and daily perfecting (appendix 4), you can synchronize with the Eternal. Then, like Jesus the Christ, you will also perceive: whoever sees you sees the Father.

Since the '**1**' is unchanging, inner peace, stillness of the mind, purity of heart, justice, wisdom, etc. (*perfections* of appendix 4) are the *modifiers* or *transformers of reality*. Akin to high-altitude hiking: the higher the altitude, the more beautiful the panorama. Similarly, the more inner peace, stillness of the mind, purity of heart, justice, wisdom, etc., the better the reality through which to experience your miracle after miracle life, liberty, and the pursuit of happiness.

A paradigm shift case for *you* to test. Fron 1986 to 2020 the number of nuclear weapons in the world has declined notably. In 1986 there were approximately seventy thousand nuclear weapons in the world. According to the Federation of American Scientists, it is estimated that in 2020 there was approximately 13,410 global nuclear warheads in inventory. Around the Earth, the equator's circumference is 24,901 miles. That is approximately one nuclear bomb every two miles. Through daily perfecting and synchronization with your inner peace, purity of heart, etc., you can reduce the global nuclear warhead inventory. Also, you will be changing your environment, your life, your liberty, your freedom, your abundance, and your happiness.

Have your Holy Family, your friends, and yourself *live* high-altitude inner peace (perfection), and before long, you will see a new Jerusalem, where like Saint John of the Cross (nicknamed "the most mystical of all poets, and the most poetic of all mystics"), a Carmelite friar of Jewish converts to Catholicism and one of the thirty-six Doctors of the Church, you will know, experience, and proclaim:

> Mine are the heavens and mine is the earth. Mine are the nations, the just are mine, and mine are the sinners. The angels are mine, and the Mother of God, and all things are mine; and God Himself is mine and for me, because Christ is mine and all for me. What do you ask for? What do you ask, then, and seek, my soul? Yours is all of this, and all is for you.

Thank you. I love you.
Orest

APPENDIX 1

Union without Ceasing

Orest has found among Jesus, the apostles and saints, and scholars and seers guidance in achieving the prayer life of union with God through love.

The world now needs prayer more than any other thing. For ages, man has been trying to find solutions to war, genocide, atrocity, poverty, slavery, murder, drug addiction, immorality, and riots, sickness, and so forth. Have we been successful? Have we come close to the real answer? No, because we have not been using the right tool: we have not been using prayer, the master key to life that our Father has put in our hands. Prayer *can* conquer all problems. Study the history of man's intercession with his Creator and you will find that when properly asked, your Father will give you whatever you need. All you have to do is ask in Jesus's name, and He will provide.

God created you and the universe, with its immensity of splendor and beauty, for you to use and enjoy. Certainly He can help you with your problems. He is waiting and longing to give you much more than the cup you hold up to be filled. But you have to turn to Him. Nobody can do it for you. You have to do it yourself. "He has knocked at the door" (Rv 3:20), and you have to open the door of your heart and let Him in. You can read about prayer. You can talk about prayer. You can listen to beautiful sermons. You can attend conferences or help with bazaars. But "let those men of zeal," recommended Saint John of the Cross, "who think by their preaching and exterior works to convert the world, consider that they would be much more pleasing to God—to say nothing of the example they give—if they would spend at least one half of their time in prayer."

Some later time, perhaps tomorrow or perhaps next year, you will take the next step of prayer, but not now. Now is not the right time. You want the gifts of God without paying the price. Which will it be for you? Will you get involved? Will you do it now? The choice is clearly yours. And the proof is in praying—constantly praying with your heart and your mind and your whole being. You will stop believing in God; you will start seeing Him! You will be consciously walking in Him. He will be your guiding light, your shining perfection, your strength, your courage, and your wisdom. *You will really feel that He dwells in you and you in Him.* Every moment will be His moment. You will drink His life. You will drink His love. You will lose yourself in Him. You and He will be one. You will cry out with a loud voice of Christ: "Eli, Eli, Lama sabachthani! which means, My God, My God, this was my destiny. I was born for this" (Matt 27:46).*

Here is a list of points you will find helpful as you grow in your prayer life:

Make a prayer a definite part of each day.
Find a quiet place.
Be comfortable.
Be attentive.
Ask God to teach you how to pray.
Know yourself.
Thank God frequently.
Pray with your heart.
Fast.
Have faith.
Be still.

Let's look at each one of these and see how they help you attain the union without ceasing, which has been called by Dr. R. M. Bucke

---

* *The Gospel Light, Comments on the Teachings of Jesus from Aramaic and Unchanged Eastern Customs,* by George Lansa © 1936, 1939, renewed © 1964 by A. J. Holman Co. Reprinted by permission of A. J. Holman Co., a division of J. B. Lippincott Co.

"cosmic consciousness," by Christian mystics "unitive life" or "beatific vision," by Buddhists "nirvana," and by Hindus "moksha."

Make prayer a definite part of each day. Set aside a certain time each day, the earlier the better. In the beginning, make it five minutes, ten minutes, or whatever your busy schedule permits. Remember: the less you pray, the less you want to pray. No one ever learned how to fly in one easy lesson. It takes time and effort. The great pianist and composer Sergei Rachmaninoff said that when he skipped one day's practice, he knew it; when he skipped two days' practice, the critics knew it; and after three days, the audience knew it. This is even truer with prayer. Every lapse is a setback and joy lost forever. The goal is to learn to pray always, in all places, without interruption.

Find a quiet place. Any place, outside or inside: an empty room, an attic, a church, a park, a garden, a lakeside bench, a mountain, a bathroom, or a closet. Be certain you will not be disturbed. True, you can pray while scrubbing pots and pans in the kitchen or while driving or on the train. However, the deeper experiences of prayer come in solitude. Jesus advised, "Whenever you pray, go to your room, close your door, and pray to your Father in private. Then your Father, who sees what no man sees, will repay you" (Matt 6:6).

Be comfortable. Traditionally Westerners kneel, because kneeling happens to be the accepted gesture of respect before an earthly throne. We know from Matthew that Jesus "threw Himself face down to the ground and prayed fervently" (Matt 26:39). Easterners seem to feel that sitting erect with spine straight and the body in equipoise is most conducive to prayer. Possibly there is no right posture. You have to pray in the position that suits you best. It's not the position of your body but the state of your heart that counts.

Be attentive. Keep your mind fixed on God. To do this, you should not be overtired, torpid, overfed, or sleepy. Your body should be tranquil and free from physical and mental tensions. The Swedes have a beautiful proverb that may be used as a guide: "Fear less, hope more; eat less, chew more; whine less, breathe more; talk less, say more; hate less, love more; and all good things are yours."

Ask God to teach you how to pray. If you think you know how to pray, listen to Saint Teresa: "Those who walk in the way of prayer have the greater need of learning; and the more spiritual they are, the greater is their need."

No matter what you have heard or read, prayer is no simple task. And it is impossible to pray properly without God's help. "Prayer is the hardest kind of work," confessed German rocket physicist and astronautics engineer Dr. Werner von Braun, "but it is the most important work we can do now." Tell Him, "Here I am, God." "Without your help, I'm sunk. Holy Spirit, take my weakness, enlighten me, and guide me into all truth." (And He will!) Let Him fill your heart and mind and kindle in you the fire of His love. Let Him make you complete—an overwhelming experience of beauty, love, and joy. Let Him bring you into the exhilarating experience of the presence of God.

Know yourself. Purify your soul and put your life in order. Search your heart for the sins you have made in thought and deed that have hurt others, as well as yourself. Realize that your sins and failures prevent God's love from working in your life, and unless you remove all barriers, the free flow of His love will be blocked. Jesus recommended, "If you bring your gift to the altar and there recall that your brother has anything against you, leave your gift at the altar, go first to be reconciled, and then come back and offer your gift" (Matt 5:23—24). No trespass should be too great for you to forgive. Hanging on the cross, the Light of Life prayed for those who were killing Him, "Father, forgive them; they do not know what they are doing" (Lk 23:34). A forgiving spirit toward others is required for effective prayer and your own health, and God will forgive you only in the measure in which you forgive. So be merciful—and set yourself free.

Often we are our own debtors. We have trespassed. We have to ask God's forgiveness. Tell Him with the heart of a child in humility and sorrow: "O Father, I am sorry! Forgive me!" And He will forgive you and forget. His forgiveness is the love that completely burns out the past sins. You then have to forgive yourself—not partially but completely. You must not recount over and over your wrongdoing. Bury the garbage of past mistakes; otherwise, it is spiritual suicide. It's like opening a wound that is once clean. You will have to make it septic again. Having confessed, turn your back upon your transgressions and be ready in the future to "hate the sin," as the jail inscription at Karnal, India, advises, but "not the sinner." Then, "avoid all evil, cherish all goodness," recommended Buddha, and "keep the mind pure." For as Saint Paul noted, you "are the temple of the living God" (2 Cor 6:16). And being holy is not only limited to deeds but also to thoughts. Jesus made this clear concerning adultery, "Anyone who looks lustfully at a woman has already committed adultery with her in his thoughts" (Matt 5:28).

All day long, thoughts that occupy your mind are molding your destiny for good or evil. Once in a vision, I saw a good and a bad thought. Chills ran up and down my spine when I realized their great power and the boomerang effect they have. Every thought, I realized then, is a high-frequency energy field that, if viewed with a suitable apparatus, would appear as real as an object you see. Nikola Tesla, for example, who invented the AC motor, transformer, radar, fluorescent light, wireless remote control, and so forth, said he could see his thoughts. They were so vivid and solid that at times it was difficult for him to distinguish between his thought and external reality. While looking at the image, he would construct many desired pieces of machinery completely in every detail and dimension. Sometimes he would test them—still in his mind—over a period of more than a year, observing the parts as they wore, modifying them, as needed.

If you could see your thoughts and their effect on your surroundings, you would be petrified to dwell on anything negative. You would realize that *the whole of your life's experience is but the outer expression of inner thoughts you have chosen to hold.* You would understand that what you think in your mind, you will invariably produce in your experience. Think love, and the love energy emanating from your mind will not only surround and modify you but all those about whom you think. Think thoughts of hate, and hate energy will be acting on you and on those about whom you think.

Dr. William Parker demonstrated in the laboratory that if you hold a feeling of hard implacability against life or impulses of unkindness or jealousy or anger or hatred (love in reverse) toward someone, you are really slowly killing yourself. It's like grabbing a hot iron to hit someone but in fact hurting yourself instead. The Man of Galilee summarized it very beautifully: "Whatsoever you sow," in your unseen thoughts, "that shall you also reap," in that which is seen. Illnesses (cancer, ulcers, heart attacks, and so on), murder, and poverty (the list is endless) are self-inflicted thoughts. You have to change the prevailing tenor of your life on love and you will see no darkness. For if you will not take the road of light, you will have to keep on learning by pain. Man cannot break the laws of God; he can only break himself against them. Only by much searching and mining will gold and diamonds be obtained. Therefore, concentrate on the right choice and true application of God-like thoughts and deeds, and you will grow through love to Love. So with every breath, with every thought, and with every aim, let love be love in you.

Thank God frequently, joyfully, and eagerly. Thank Him with the deepest affection of your heart for letting you be alive and part of His creation. Praise His goodness and love for you. Recall the many graces of mercy, happiness, joy, wisdom, ability, and health He has given you; the many sins He has forgiven you; and your ability to see His glory and drink from the spring of His Spirit. Thank Him always for everything that happens to you during every moment of your life.

"We don't thank God enough for much that He has given us," confessed Robert Woods. "Our prayers are too often the beggar's prayer, the prayer that asks for something. We offer too few prayers of thanksgiving and praise." Remember when Jesus healed the ten lepers? Why did only one return to give thanks to God? Were ten not cleansed (Lk 17:14–19)?

"When I look at your heavens," cried David in Psalms 8:3–4, which "the work of thy fingers, the moon and the stars you have established; what is man that thou art mindful of Him?" We are so small, our life so short, and our knowledge so limited, we begin to feel with Tennyson that the life of men is but "murmur of gnats in the gleam of a million suns." Thank God; adore Him because "Thou has made us for Thyself," sang Saint Augustine, "and our hearts can find no rest outside of Thee."

Pray with your heart. When you pray, realize that you are addressing the Maker of all. He is the love of loves and the Creator who is infinitely bigger to our universe than the universe is to an atom—the absolute beauty, gentleness, wisdom, goodness, and power. Do not address Him while you are thinking of other things. Show your gratitude to Him for allowing you to come near Him. Remember, before you can really pray with sincerity you must begin to know the greatness and holiness of God. And the greater the depth and extent of the knowledge, the more love there will be. The more easily the heart will soften and lay itself open to the love of God.

God knows before you open your lips what you will say. You must pray with your mind, your heart, and your whole life. You must seek to know His will, be eager to offer yourself to Him, and be ready to be filled with His Holy Spirit. You must pray in your own words, just the way you talk to friends. The cry from the heart of the most illiterate is just as welcome to God as the perfectly formed prayer of a great scholar. But it is not the utterances of the lips that God hears; rather, it is the song and joy of the heart. If your heart does not speak, you are silent to God. Love God and He will hear you.

Fast. Deliberately abstain from food for spiritual purposes. Why fast? Isn't a good life adequate? Why did Jesus fast? Why did Moses, Samuel, David, Elijah, Daniel, Isaiah, Cornelius, Paul, Socrates, Plato, Buddha, Gandhi, Pythagoras, Luther, and Lincoln fast? Did they know something we don't know? Christ did not have to fast! He is the Light of Life and Perfection. Yet He taught, fasted, performed miracles, was crucified, died, and arose not for exhibition but to teach us how to grow. In His Sermon on the Mount, Jesus did not say *"if* you pray . . . *if* you fast . . . *if* you give alms" but *"when* you pray . . . *when* you fast . . . *when* you give alms." His language for prayer and fast is identical. He expects us unambiguously and without qualification to pray, to fast, and to give when the occasion demands it. There is one requirement, however: when you fast, please God and not the eyes of other people. "Appear not to men that you are fasting, but to your Father who is unseen: and your Father, who sees what is hidden will reward you" (Matt 6:18). The shameful hypocrisy, the egocentric piety, and showy acts of fasting of the Pharisees had no place in Jesus's life and must be curbed in your life. Fasting should be inconspicuous, noncompetitive, and uninjurious to your health. A fast of one to three days is easy and does not fall into this last category. But if you embark on a longer fast (seven to forty days) be sure that God is leading you to do this and that you understand how to go about it. If you have an illness or doubts as to the physical advisability of fasting, consult your own doctor and follow his advice.

So why are you afraid to fast? Why are you not doing what Jesus has instructed us to do? Very simply: the evil spirits (earthbound, unenlightened individuals) that Christ had to deal with are working against you. Every opportunity the dark forces get or every roadblock they can set up (chaos, disruption, fear, hatred, worry, aggression, or separation) goes into action. Fasting clears the doors of perception and removes the impediments to holiness. It helps to crucify self-love and self-will and allows God's overwhelming power to shape the real, eternal you. It breaks down that which stands between God and you and *tunes your mind and body to be a better instrument for love.*

Both the Old and the New Testament very clearly state (seventy-four times) that prayer and fasting are a must, that "man cannot live by bread alone" (Matt 4:4) and that some evil spirits can only be driven out by prayer and fasting (Matt 17:21). People who have fasted will testify that it constitutes one of the most powerful tools God has put in your hands. When you fast you bring a note of urgency to your prayer. You are telling God that you are truly in earnest, that you do not intend to take no for an answer, and that you want a miracle.

The great prophet Isaiah, who vividly foretold the suffering and death of Jesus, listed some of the benefits of fasting: answer to prayer, health, guidance, healing, inspiration, and the Glory of the Lord shall be your reward (Is 58:1–14). Can you wish for more? Individuals who fast claim many other benefits: Daniel—improvement of his prophetic ability; Elijah—spiritual direction; Socrates, Plato, Plutarch, Pythagoras, and Galileo—sharpening of intellect and mental clarity; the people of Israel—divine intervention; the apostles—spiritual enrichment; Krishna, Buddha, Shankara, Confucius, Gandhi—understanding of truth; and the eight thousand (1948 through the present) patients under Dr. Yurij Nikolayev at Moscow's Gannushkin Institute—treatment of sluggish forms of schizophrenia.

The famous physical fitness authority Dr. Paul Bragg testifies that "you purify your body physically, mentally, and spiritually and therefore enjoy super vitality and super health. Greatest of all are the inner peacefulness and tranquility that make life worth living. You come into harmony with that power higher than yourself. You learn the meaning of the truth that "your body is the temple of the living God." How to start? Very simply:

You have to believe that fasting is good for you.

If this is your first, don't start on a long fast. Learn to walk before you run.

Have fruit for your last meal.

Abstain completely (initially for twenty-four hours) from eating and drinking everything except water.

Drink as much boiled, warm water as you wish.

The first time you fast you may get a headache, especially if you are used to drinking coffee, tea, or alcohol. This is one of the signs that your body is undergoing detoxification. It is unpleasant but good for you. You might consider stopping coffee, tea, or alcohol a few days before the fast.

To fight discouragement (evil spirit attacks), praise God.

Break fast with fruit or vegetable juices (tomato or citrus).

At the following mealtime, have fresh salad (without dressing) or homemade vegetable soup (no fat). Avoid pastries, biscuits, and starchy foods.

At first sensation of fullness stop eating.

To avoid pain and discomfort after a longer fast, Dr. Herbert Shelton (who has supervised over thirty thousand healing fasts) advises that after the fast, you discipline yourself and control your appetite and food intake.

Have faith that God will solve your problems. Realize that all things come from Him out of His love for us. "Ask, and you will receive," recommended the Man of Nazareth. "Seek, and you will find. Knock, and it will be opened to you. For the one who asks, receives. The one who seeks, finds. The one who knocks, enters" (Mt 7:7–8). Jesus's words assure you that the door is wide open to receive whatsoever you desire for yourself or your loved ones. Just lay before Him the needs of your soul and body, *believing that you have already received them.* Open your heart reverently and wholeheartedly, and He will bestow the richest blessing. "Would one of you hand his son a stone when he asks for a loaf, or a poisonous snake when he asks for a fish?" further taught Jesus. "If you, with all your sins, know how to give your children what is good, how much more will your Heavenly Father give good things to anyone who asks Him" (Mt 7:9–11).

Your Father, the ultimate spring of living water, is longing to help you with your needs. Therefore, *be always alert and attuned to Him for inner guidance, direction, and support.* But once you know what has to be done, you have to get up and do it. Suppose your house were in darkness, yet there is a power line directly outside. All you have to do is bring that electric line into your home. You have a choice: you can curse the darkness or you can link the Eternal Light into your life. However, if you assume that the purpose of prayer is only to get what you want in material things from God, then you may never rise far. There is a deeper purpose and meaning to prayer. Pray with humility for spiritual strength, help in avoiding sin, guidance, wisdom, understanding, and love. Pray for others, and hold them up into the light of God's presence.

"There is nothing that makes us love a man so much as praying for Him," confessed William Law. "By considering yourself as an advocate with God for your neighbors and acquaintances, you would never cease to be at peace with them yourself . . . such prayers as these amongst neighbors and acquaintances would unite them to one another in the strongest bonds of love and tenderness." Pray for those who hate you; pray for those who persecute you. Pray, love, and serve men in utter selflessness. Be a channel through which God may act. We do not live alone, and we do not die alone. "We are members of one another" (Eph 4:25); "I am the vine, you are the branches" (Jn 15:5), taught the Light of Life. *Everything we say or do has some influence on everyone in the human community.*

The best petition is not to reach out in your own way for what you don't understand completely but to leave yourself in the arms of your Father and *let His will be your will.* Let His law be your law. Let "not I, but Christ in me" be your motto. Remember, when Jesus prayed in the garden of Gethsemane, He asked to be relieved of the cross, "Abba [oh, Father], you have the power to do all things. Take this cup away from me. But *let it be as you would have it, not as I*" (Mk 14:36).

Be still, and listen to God within. If you were with a king or a president, would you be talking all the time? Would you be asking without stopping—especially if the ruler could read your mind and know your heart's desires before you opened your lips? "Your Father," said Jesus, "knows what you need before you ask Him" (Matt 6:8).

Having done your talking and petitioning, wait before Him in love, joy, adoration, and devotion and listen to Him.

God speaks to your heart in silence. Silence means more than ceasing to speak with your lips; it also means practicing and *maintaining stillness in your mind* (when you turn within, leave the problem outside). Sometimes when you are praying, you may not speak at all with your lips, but your mind is boiling over with emotions and fears so that you cannot hear the "still small voice" of God within you.

Dom John Chapman captured the essence of silence when he stated that "you can't make silence—you can make noise. But you can only make silence by stopping the noise." The prophet Habakkuk testified, "But the Lord is in His holy temple: let all the earth keep silence before Him" (Hab 2:20). We should not let our cries for our earthly needs disturb the inner peace of God's temple. "A man does not see himself in running water," confirmed the sage Chuang Tzu, "but in still water." So make the stillness your own. Turn your creative faculties into a receiving station and start listening not only with your ears but with your entire being to the still, small voice of God's whisper. As you listen and rest in His love, you may come to feel as though the whole world were vibrating with the presence and love of God with absolute peace and stillness and yet with an intense and ceaseless energy. *In this stillness, you will learn who you really are.* You will learn of love. You will learn of patience, humility, peace, joy, and life. You will learn of Him. This learning you have to do yourself. No book or teacher can tell you the feeling of quietness, clearness, stillness, love, or beauty. You have to experience it yourself. There is no other way. The teacher can show you the direction, but you have to throw your whole self into the journey. You have to "be still and know that I am God" (Ps 46:11).

An unknown writer expressed, "Wait still upon God. Open your heart to Him, let the light and warmth of His love flood your mind and heart and soul as silently as the flower opens itself to, and drinks in, the light and warmth of the sun, and becomes itself truly beautiful, and thereby rest in the conscious thought of your living union with Christ."

The eleven points discussed above have one purpose: to help us achieve *union without ceasing. Webster* defines *union* as "an act of joining together, or a state of being united." What actually takes place, said Saint Teresa of Avila, is that "your soul becomes one with God." Saint

Paul defined union in similar way: "Whoever is joined to the Lord becomes one spirit with Him" (1 Cor 6:17). The great Indian (Hindu) philosopher Radhakrishnan elaborates further, "The oldest wisdom in the world tells us that we can consciously unite with the divine while in this body, for this is man really born. If he misses his destiny Nature is not in a hurry; she will catch him someday and compel him to fulfill her secret purpose."

We know that Our Father is within. He is also without. Just like all the TV and radio programs. To hear or see them we need a tuner. *That tuner to our Father is* purity (in thought, word, and deed), *stillness, and love.* Incidentally, if you are not perfectly aligned on your radio or TV dial, you will also be getting noise. Your program will be disturbed. The same is true here. *Your mind has to be emptied of sensations, images, and thoughts.* You have to forget yourself. You have to concentrate on the love of God and the Light within your soul, all universe, and beyond. *Be still* and *love Him* with all your heart, *be still* and *love Him* with all your soul, *be still* and *love Him* with all your power, *be still* and *love Him* with all your mind (lose yourself in the Beloved), and nothing else.

In this loving embrace and bliss, your spirit will be absorbed into eternal love. All will become one, like one light merged in the ocean of light. You will transcend the uttermost bounds of anticipation or desire. You will have reached a fountain of holy joy and peace so overpowering that it transcends all other joys and passes all understanding. You will have returned to your Father's heart. Similarly, you may also concentrate on His will. Let the Father's will permeate all of you. Let Him make you complete. Let Him exhilarate you. Let Him make you one. In this splendor of joy you will notice that you as such have disappeared. Only His will, His love, and His light has remained.

Initially, the state of rapturous union can be momentary, with exhilarating joy and exultation—later on of longer durations (lasting for hours) and extending even into sleep. And eventually you can establish a permanent conscious (living) union with God. The impact of each experience is most overpowering. It "penetrates to the very marrow of your bones," testified Saint Teresa. "Your senses could be fused into one ineffable act of perception. Differences between time, space, and motion will cease to exist. You will understand the profound truth *that there is only God*, the I Am Who Am" (Ex 3:14), that "the Father and I are one" (Jn 10:30), that "I am in the Father and the Father is in me" (Jn 14:11), that "the life I live now is not my own; Christ is living in me" (Gal 2:20), that "My Me is God, nor do I recognize any other Me except my God Himself" (Saint Catherine of Genoa), and that "whoever has seen me has seen the Father" (Jn 14:9).

You may find yourself in the center of stillness and living glow so pronounced that the distinction between you and your surroundings will disappear. You will see that you and the rest of the surroundings are one and the same light and stillness, one and the same joy of shimmering, conscious glory and love. You will be moving, sitting, and working in it. It is a fantastic and indescribable splendor of delightful stillness and peace. The mind thinks—but for some reason, all of you ceases to exist as a separate entity but becomes a part of one infinite light and love. You may dive into an ocean of knowledge (truth, basic concepts of science, works of art, or inventions) where all that was obscured is now explained, where all problems are solved, and all that is or will be knowable is known. All knowledge of that which is above all reason and beyond all thought is nearer to you than you are to yourself. You have tuned in to your Father's heart.

"This is the way," said the voice of God to Saint Catherine of Siena. "If you will arrive at a perfect knowledge and enjoyment of Me, the Eternal Truth, you should never go outside the knowledge of yourself; and by humbling yourself in the valley of humility you will know Me and yourself, from which knowledge you will draw all that is necessary." Saint Teresa of Avila, who drank God's wine of union, similarly related, "There will suddenly come to it [soul] a suspension in which the Lord communicates most secret things, which it seems to see within God Himself. . . . The brilliance of this vision is like that of infused light or of a sun covered with some material of the transparency of a diamond . . . For as long as such a soul is in this state, it can neither see nor hear nor understand: the period is always short and seems to the soul even shorter than it really is. God implants Himself in the interior of that soul in such a way that, when it returns to itself, it cannot possibly doubt that God has been in it and it has been in God; so firmly does this truth remain within it that, although for years God may never grant it that favor again, it can neither forget it nor doubt that it has received it."

Saint Catherine of Genoa, who was one of the most penetrating gazers into the secrets of eternal light, stated, "When the loving kindness of God calls a soul from the world, He finds it full of vices and sins; and first He gives it an instinct for virtue, and then urges it to perfection, and then by infused grace leads it to true self-naughting, and at last to true transformation. And this noteworthy order serves God to lead the soul along the Way; but when the soul is naughted and transformed, then of herself she neither works nor speaks nor wills, nor feels nor hears nor understands, neither has she of herself the feeling of outward or inward, where she may move. And in all things it is God who rules and guides her, without the mediation of any creature. And the state of this soul is then a feeling of such utter peace and tranquility that it seems to her that her heart, and her bodily being, and all both within and without is immersed in an ocean of utmost peace; from when she shall never come forth for anything that can befall her in this life. And she stays immovable, imperturbable, impassable. So much so, that it seems to her in her human and her spiritual nature, both within and without, she can feel no other thing than sweetest peace. And she is so full of peace that though she press her flesh, her nerves, her bones, no other thing comes forth from them than peace."

Having immersed yourself in perfect love for a new life and a new purpose, you can't stop here. You have to go on. You have to bring forth your fruits in good deeds. You have to serve others. You have to *treat every person as you would treat Jesus.* See Him in every heart and every face. For He said it Himself, and we shall hear it again in a day of judgment: "I assure you, as often as you did it for one of my least brothers, you did it for me" (Matt 25:40). Say nothing, do nothing, and think nothing that is not love-directed. If what you do—be it in thought, word, or deed—does not create love, don't do it. Remember, you are not just rendering it to a mortal man but unto God who is within that man.

A note of caution: As you are journeying onward and upward to God (growing in goodness and love), you may come upon many "wine cellars," "fireworks," and "beautiful sceneries" full of surprises. As the Bible shows (1 Cor 12:9, 10; 2 Cor 12:1), these by-products could exhibit themselves in many forms:

You may speak or interpret tongues.

You may experience visions and revelations, prophecy, and have the power to distinguish one spirit from another.

You may acquire a new kind of perception where the whole cosmic panorama may appear magnified and full of light and grandeur.

Your body sweat may emanate perfume-like fragrance.

You may acquire the healing touch.

You may exhibit bilocation (go to distant places with your body instantaneously (Jn 20:26–9; Act 8:39–40). This is possible because as you concentrate on God, His forces dematerialize the body. It becomes less solid and more flexible (your body frequencies increase) and therefore can be more easily acted on by your thought. For additional reading, read about people who had this quality: Saint Ignatius, Saint Clement, Saint Francis of Assisi, Saint Anthony of Padua, Saint Francis Xavier, Joseph Cupertino, Saint Martin de Porres, Saint Alphonsus Liguori, and Padre Pio.

These and similar diversions may so overpower you that you will become arrested at this level of growth. True, the dramatic phenomenon is fascinating, but it should *never* become for you the circumference of your horizon. Similarly, it is highly dangerous to one's soul and health to seek gifts of this nature for their own sake or for personal enhancement. Let the scenery be there (use it in a constructive way), *but keep in mind your destination is the living union with Your Heavenly Father* and nothing less. And at the union, you have to arrive not by expending consciousness with drugs, chemicals, and hallucinogens, but through the way of Jesus. With drugs and chemicals, you are not improving yourself, you are drowning yourself. You are letting yourself be possessed by saboteur and insidious spirit entities, with results that are far worse after the trip than before the trip.

Listen to what the Son of Man had to say about this: "I tell you the truth: whoever does not enter the sheepfold by the door, but climbs in some other way, is a thief and a marauder. To get there," said He, "I am the door. Whoever enters through me will be safe. He will go in and out and find pasture" (Jn 10:9). Similarly, Jesus reproached the Pharisees, most of whom were ignorant of union yet deliberately obstructed the helpless masses: "Woe to you lawyers! You have taken away the key of knowledge. You yourselves have not gained access, yet you have stopped those who wish to enter!" (Lk 11:52). The key to union you will find only through crucifying your self-love and self-will. Jesus demonstrated this with His death on the cross and demanded that we do the same: "Whoever wishes to be my follower must *deny his very self*, take up his cross each day" (not the cross of suffering, but the self-crucifixion) and "and follow in my steps" (Lk 9:23).

True, the lower self doesn't want to die. It wants to live; it wants Father to take that cup away, but your conscious self knows that "for he who wishes to save his life must lose it." He must destroy the lower self so that it will not be the deciding guide, but the higher will, the Father's will, will be the way. You have to consciously live in God's presence, a life of Jesus Christ (make iron determination to be pure within), a life of "be perfect just as your Father in heaven is perfect" (perfect in love), continuously a life of *love in action* (love everything because it needs love like you; your Father has programmed "the branches" that way, so "that all may be one . . . as we are one") . . . one day at a time—now. The reward of this type of life is union without ceasing, where Eternal Love and you will be one; where heaven and all glory will exist here and now; where the state of joy, splendor, bliss, tranquility, and overpowering peace will be with you every moment; where with tears in your eyes, you the prodigal son will be embraced by your Beloved Father on your welcome home; and where, with a might of Christ, you will proclaim: "Eli, Eli, lmanashabachthani! . . . My God, My God, this was my destiny. I was born for this."

Then you will know this chapter is finished. Now a new chapter of resurrection, light, and unity is upon you.

I Love You!

Basic Sources

Biblical references in italics are from the *New American Bible;* others are from *Good News for Modern Man.*

The New American Bible, © 1970 by Confraternity of Christian Doctrine (The Catholic Press, Washington DC)
Good News for Modern Man
Today's Version of the New Testament, © 1966, 1971, American Bible Society

APPENDIX 4

UDC 517.9:519.46
© 1993

O. BEDRIJ

SCALE INVARIANCE, UNIFYING PRINCIPLE, ORDER AND SEQUENCE OF PHYSICAL QUANTITIES, AND FUNDAMENTAL CONSTANTS

(Submitted by Corresponding Member W. I. Fushchich)

Scale invariance is a fundamental concept in physics. The concept of scale invariance can be expanded to enable us to more fully articulate fundamental questions concerning the scale, sequence, and classification between different dimensional quantities. Scale invariance permits us to shed new light on the innermost structure, the deep and profound order, and the mutual interaction of the fundamental constituents of physics. Also, scale invariance helps us derive new laws of nature and fundamental physical constants.

When relationships between different physical quantities are expressed by an equation with an equal sign (=), it means that the corresponding physical quantities are in equilibrium. For example,

$$q_1 = q_2 q_3, \text{ or } 1 = q_1/q_2 q_3 = q_2 q_3/q_1, \ q_1 \neq 0, \ q_2 q_3 \neq 0, \tag{1}$$

where $q_1 = V$ (electric potential in volts), $q_2 = R$ (resistance in ohms), $q_3 = i$ (magnetic potential in amperes). Formula (1) is the Ohm's law, which shows that q_1 is in equilibrium with $q_2 q_3$. Formula (1) is a simple expression of a more general form:

$$1 = (q_1^{x_1} \cdot q_2^{x_2} \cdot q_3^{x_3} \ldots q_s^{xs}) / (p_1^{j_1} \cdot p_2^{j_2} \cdot p_3^{j_3} \ldots p_z^{jz}) \tag{2}$$

or

$$1 = Y'/KX, \tag{3}$$

where

$$Y' \int (q_1^{x} \cdot q_2^{x} \cdot q_3^{x} \ldots q_s^{x}), \tag{4}$$

$$1/KX \int (p_1^{j} \cdot p_2^{j} \cdot p_3^{j} \ldots p_z^{j}), \tag{5}$$

$$(q_s)^0 = 1, \; (q_s^{-1})^0 = 1,$$

$q_1, q_2, q_3, \ldots, q_s, p_1, p_2, p_3, \ldots, p_z$ are quantities, $x_1, x_2, x_3, \ldots, x_s, j_1, j_2, j_3, \ldots, j_z$ are real numbers, K is the slope for line $Y' = KX$, $j, s, x, z = 1, 2, 3 \ldots$ We require that formula (2) is scale invariant. That is, formula (2) is invariant with respect to the following transformations:

$$q_1 \rightarrow q_1' = aq_1, \;\; q_2 \rightarrow q_2' = aq_2, \;\; q_3 \rightarrow q_3' = aq_3, \ldots, \tag{6}$$

$$p_1 \rightarrow p_1' = ap_1, \;\; p_2 \rightarrow p_2' = ap_2, \;\; p_3 \rightarrow p_3' = ap_3, \ldots \tag{7}$$

Where «a» is scale transformation parameter, and all physical quantities (q_s and p_z) have to be subject to transformation. Hence, based on formula (2), it follows that «1» is always invariant with respect to scale transformation (6) and (7).

When a structure of physical relationships requires a more complete description, we can expand formula (2) to the following formulae:

$$1 = Y'/(KX + C'), \tag{8}$$
$$1 = Y'/(B'X^2 + KX + C'), \tag{9}$$
$$1 = Y'/(D'X^3 + B'X^2 + KX + C'), \tag{10}$$
$$1 = Y^2/(B'X^2 + KX + C'), \tag{11}$$
$$1 = (A_2 Y^2 + A_1 Y')/(B'X^2 + KX + C'), \tag{12}$$

where K, C', B', D', A_1, A_2 are real parameters.

Please note that every physical quantity of equation (2) has its own dimension that, when cross multiplied in (2), becomes dimensionless.

Indeed, we have a relationship for each physical quantity as it relates to all other quantities in (2). Further, we can consider all these relationships as algebraic equations to define the relative dimensionless value of each physical quantity.

For example, let us consider equation (2) and the following relationships:

$$1 = q_1/q_2 \cdot q_3 = q_2 \cdot q_4 = q_5 \cdot q_6 = q_7 \cdot q_8, \qquad (13)$$

where $q_4 = G$ (conductance in siemens), $q_5 = T$ (period of harmonic motion in seconds/cycle), $q_6 = f$ (frequency of the motion in cycles/second), $q_7 = \lambda$ (wavelength in meters/cycle), $q_8 = n$ (wave number in waves or cycles/meter).

From formula (2) or (13) we can see that in cross multiplication, dimensions disappear. Then relationships (2) or (13) can be considered as an algebraic relation for dimensionless quantities, with «1» as the central pivot and the least common denominator of all physical relationships (2)-(13). Further, «1» can also be viewed as a product of quantity and its reciprocal ($q_5 q_6 = q_7 q_8 = 1$).

From formulae (1), (2), and (8)-(13), we see that «1» can be viewed as the ultimate unifying principle (Absolute) between all physical quantities and relationships of type (2). Further, that «1» is the scale invariant equilibrium frame of reference between every physical relationship. In the logarithmic or natural log scale (see Table 1), the equilibrium frame of reference «1»→0 ($1 = 10^0 = e^0$). Therefore, as in numbers, so in physics, we can consider zero the starting frame of reference for the scale of all physical relationships.

The concept of «0» (zero) is well-known and understood in mathematics. However, in physics, the concept of zero has not been adequately defined mathematically or discussed as it relates to the scale of invariance for the laws (quantity relationships) of physics and the fundamental physical constants. Even in the vacuum idea in quantum field theory, «zero» is not a complete zero concept because the quantum vacuum exists in time and space.

In numbers, while direct understanding of the zero concept came into being only recently (approximately in the ninth century), its implied value was always a part of every rational number. That is, the frame of reference of every number on a mathematical scale is «0» (zero).

We propose that just as each number, on a mathematical scale, has a unifying principle (zero) as its starting frame of reference, so each physical quantity, on the physics scale of quantities, also has a unifying principle (equilibrium frame of reference—zero) as its starting frame of reference. Further, «1» includes within itself all physical quantities and serves as the ultimate principle out of which all physical quantities and relationships (reality) emerge (2)-(13).

Equation (2) gives us only relationships between the «1» and physical quantities. To locate those quantities on one scale, we have to know

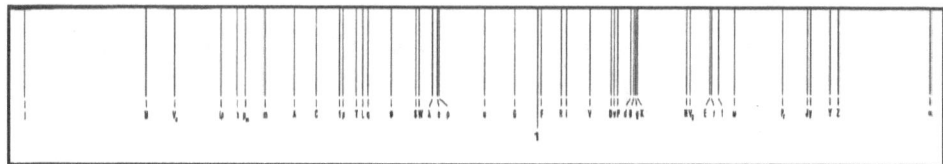

Figure 1: Spectrum of Physical Quantities

their absolute values (constants) with respect to the equilibrium frame of reference. Further, all relationships between the constants (numbers) on the «1» scale must be scale invariant. Hence, the whole spectrum of numbers must have one-to-one correspondence between physical quantities. The challenge is to find such numbers that when cross multiplied by themselves will satisfy formulae (2) and (13). Further, when the laws of physics (physical relationships) are involved, we must find numbers that will satisfy all relationships for all physical quantities (2) that $q_1, q_2, \ldots, q_8$ and $p_1, p_2, \ldots, p_8$ are part of.

At the present, we know of over seventy physical quantities and over three hundred physical relationships that the physics world has directly or indirectly verified. The numbers (constants) that must be found for formula type (2) or (13) must satisfy simultaneously all these relationships in three hundred equations. That is, we must find constants that will be applicable to all quantities of type (2) that have been discovered and verified by physics.

To solve the problem, we have used computers. [1, 2] When we performed the computations, we found that fundamental constants such as Planck's constant, velocity of light, electron charge and mass, Compton wavelength, permittivity and permeability of free space, Rydberg, and the fine-structure constant enter into all physical relations (laws of physics). We also found additional (new) physical relationships and constants. A partial list of the computed constants is shown in Table 1.

When we affix appropriate numbers (constants) with their corresponding physical quantities, from Table 1, on a scale of one axis, we obtain a spectrum of physical quantities—a natural scale of measurement with its fixed order and sequence of physical quantities, with «1» in the center of the spectrum (Figure 1). The spectrum has a colossal range. For example, when we use the electron constant as the frame of reference, the range will extend from 10^{-57} for the moment of inertia to greater than 10^{43} for angular acceleration. [2]

By combining two physical quantity spectra (two Figure 1s), we can construct a logarithmic slide rule of physical relationships (LSPR). Instead of the numbers on a presently used slide rule, we have physical quantities with LSPR. As with numbers that have a sequence on a slide rule, with LSPR we have a sequence of physical quantities. With a mathematical slide rule we can multiply and divide numbers; with an LSPR we can multiply and divide physical quantities.

Transformations from one physical quantity to another are achieved by shifting (translating) the frame of reference, of 10^0 power, to another frame of reference, having a new equilibrium center of 10^n or 10^{-n} power. Note that all fundamental laws of physics are in equational equilibrium. That is, the left side of the equation equals the right side of the equation. Just as the equal sign allows one to go from one system of measurement to another, so «1» on the LSPR allows one to shift the equilibrium center from one frame of reference (quantity) to another. The equilibrium center defines the frame of reference.

Table 1. Fundamental Constants of Quantum Electrodynamics

Symbols	Constants	Natural Log	Relationships of Quantities
I	$1.142280538 \cdot 10^{-57}$	-131.1143236	$I = mS^2$
V_0	$4.440432224 \cdot 10^{-41}$	-92.91523709	$V_0 = m/d$
h	$6.626075438 \cdot 10^{-34}$	-76.39388047	$h = W/f$
m	$9.109389672 \cdot 10^{-31}$	-69.17083217	$m = F/y = V_0 d$
A	$1.253959463 \cdot 10^{-27}$	-61.9434914	$A = \Phi/B = i/J$
C	$3.135382131 \cdot 10^{-25}$	-56.42187627	$C = q/V = q^2/W$
u	$1.700877491 \cdot 10^{-24}$	-54.73089794	$u = iA = W/B$
$t\ (1/\omega)$	$1.181193493 \cdot 10^{-22}$	-50.49034668	$t = q/i = p/F$

$T\,(1/f)$	$8.093300952 \cdot 10^{-21}$	-46.26325028	$T = 1/f = h/W$
$L\,(1/r)$	$4.449913947 \cdot 10^{-20}$	-44.5588171	$L = \Phi/i = S\mu$
q	$1.60217733 \cdot 10^{-19}$	-43.27775323	$q = CV = F/E$
Q	$3.759275896 \cdot 10^{-19}$	-42.42489041	$Q = Av - V_0 t$
Φ	$6.035887677 \cdot 10^{-17}$	-37.34622365	$\Phi = F/H = BA$
S	$3.5441129005 \cdot 10^{-14}$	-30.97174576	$S = V/E$
W	$8.187111141 \cdot 10^{-14}$	-30.1336302	$W = Pt = FS$
$\lambda\,(1/n)$	$2.426310586 \cdot 10^{-12}$	-26.74464929	$\lambda = v/f$
ε	$8.854187818 \cdot 10^{-12}$	-25.45013057	$\varepsilon = D/E = C/S$
Q_m	$7.712021552 \cdot 10^{-09}$	-18.68048548	$Q_m = Qd = m/t$
μ	$1.256637061 \cdot 10^{-06}$	-13.5870714	$\mu = B/H$
M	$7.264500922 \cdot 10^{-04}$	-7.227340774	$M = m/A$
$G\,(1/R)$	$2.654418727 \cdot 10^{-03}$	-5.931529583	$G = i/V = Cf$
α_f	$7.297353079 \cdot 10^{-03}$	-4.920243588	$\alpha_f = S/2\lambda = 1/2\theta$
«1»	$1.000000000 \cdot 10^{0}$	0.000000000	«1» $= GR = Tf$
F	$2.312005898 \cdot 10^{0}$	0.8381155014	$F = qE$
θ	$6.851799475 \cdot 10^{1}$	4.227096408	$\theta = \lambda/S = \omega/f$
$R\,(1/G)$	$3.767303134 \cdot 10^{2}$	5.931529583	$R = V/I$
i	$1.356405483 \cdot 10^{3}$	7.212593453	$i = P/V$
η	$2.177842587 \cdot 10^{5}$	12.29126021	$\eta = W/Q = EH/Y$
V	$5.109990628 \cdot 10^{5}$	13.14412304	$V = W/q$
R_∞	$1.097373153 \cdot 10^{7}$	16.21101493	$R_\infty = \alpha^3{}_f/S$
D	$1.277694676 \cdot 10^{8}$	18.66573816	$D = q/A$
V	$2.99792458 \cdot 10^{8}$	19.51860099	$v = H/D = E/B$
P	$6.931219308 \cdot 10^{8}$	20.35671649	$P = iV = Fv$
D	$2.051464635 \cdot 10^{10}$	23.74440492	$d = m/V_0 = P_r/V_g$
B	$4.813463159 \cdot 10^{10}$	24.59726775	$B = E/v$
G	$7.495967321 \cdot 10^{10}$	25.04021612	$g = H/V = G/S$
$\Lambda\,(1/S)$	$2.823958118 \cdot 10^{13}$	30.97174576	$\ddot{E} = E/V$
H	$3.830432276 \cdot 10^{16}$	38.18433915	$H = i/S = F/\Phi$
V_g	$8.987551787 \cdot 10^{16}$	39.03720197	$V_g = W/m$
E	$1.443039952 \cdot 10^{19}$	44.11586873	$E = V/S = F/q$
$r\,(1/L)$	$2.247234468 \cdot 10^{19}$	44.5588171	$r = i/\Phi$
$f\,(1/T)$	$1.235589787 \cdot 10^{20}$	46.26325028	$f = W/h = v/\lambda$

ω $(1/t)$	$8.466013454 \cdot 10^{21}$	50.49034668	$\omega = f\theta$
P_r	$1.843764464 \cdot 10^{27}$	62.7816069	$P_r = F/A = BH$
J	$1.081698032 \cdot 10^{30}$	69.15608485	$J = i/A$
Y	$2.538046983 \cdot 10^{30}$	70.00894767	$Y = F/M = v\omega$
Y	$4.075084387 \cdot 10^{32}$	75.08761443	$y = V/A$
Z	$3.493766483 \cdot 10^{33}$	77.23628844	$Z = \alpha/d$
S_p	$5.527466807 \cdot 10^{35}$	82.30020788	$S_p = P/A = EH$
α	$7.167338381 \cdot 10^{43}$	100.9806934	$\alpha = W/I = 1/LC$

Because not all physical relationships between quantities are presently known, with LSPR we can obtain new (unknown) relationships between physical quantities, as well as find new fundamental physical constants. Some of the new physical relationships are shown in equations (14)-(26). For explanation of terms, refer to the definition of quantities below:

$$S = a_f^3/R_\infty = (q/D)^{1/2} = (i/J)^{1/2}, \tag{14}$$

$$\lambda = \alpha_y^2/2R_\infty = S/2\alpha_f, \tag{15}$$

$$\alpha_f = 1/2\theta = S/2\lambda = tf/2 = (SR_\infty)^{1/3}, \tag{16}$$

$$\theta = \lambda/S = \omega/f = T/t, \tag{17}$$

$$L = S\mu = B/J = \alpha_f^3 u/R_\infty, \tag{18}$$

$$C = S\varepsilon \tag{19}$$

$$Y = S/LC = S\alpha = (P_y Z)^{1/2} = (V_g \alpha)^{1/2} = R_\infty V_g/\alpha_f^3, \tag{20}$$

$$h = q\Phi\theta = V_0 P_r/f = \Phi i/f, \tag{21}$$

$$m = \mu q^2/S, \tag{22}$$

$$V_g = YS = A/LC, \tag{23}$$

$$\alpha = 1/LC, \tag{24}$$

$$d = DE/V_g = BH/V_g, \tag{25}$$

$$\eta = EH/Y = W/Q, \tag{26}$$

Note: One of the key elements of LSPR is that it shows visually how all physical quantities are interrelated and classified among themselves and how the same essence—the same reality—appears in different forms. Further, the different forms—physical quantities and constants—once decoded and classified, have a very orderly and logical structure. For example, it is well-known that when an area bounded by a curve is rotated about an axis, a volume is produced. Similarly, other physical quantities can be considered in this way and predicted from physical relationships. Listed below are a few examples from the left side of LSPR:

Volume (V_0)	= Angular area	($A \times S$),	(27)
Capacitance (C)	= Angular permittivity	($\varepsilon \times S$),	(28)
Self inductance (L)	= Angular permeability	($\mu \times S$),	(29)
Electric dipole moment (e)	= Angular electric flux	($q \times S$),	(30)
Magnetic dipole moment (b)	= Angular magnetic flux	($\Phi \times S$).	(31)

Please note that all physical quantities on the left side of LSPR are complementary to those on the right side of LSPR. Hence, to produce an angular quantity, when going from the left side of LSPR to the right side, instead of multiplication by S, we would divide the quantity by S. Listed below are a few examples from the right side of LSPR:

Left Side of LSPR

Voltage density (Y)	= Angular electric field strength (E/S),		(32)
Current density (J)	= Angular magnetic field strength (H/S),		(33)
Mass Potential Density ($á$)	= Angular acceleration (y/S),		(34)
Electric Field strength (E)	= Angular electric potential (V/S),		(35)

Magnetic field strength (H) = Angular magnetic potential (i/S), (36)

Mass field strength
(acceleration) (Y)
= Angular mass (gravitational),
Potential (V_g/S). (37)

Please notice that as a rotational line produces an area, and a rotational area produces volume, in the same way the rotation of electric and magnetic potential produces electric and magnetic field strength (35), (36), and rotation of electric and magnetic field strength produces voltage and current density (32), (33). Because of space limitations, we are unable to describe the LSPR more in detail.

Summary: Views and appraisal. Using scale invariance as a fundamental concept in physics, we propose an approach for the integration of a scattered and immense body of fundamental physical phenomena into a more systematic order. The approach permits analysis of every corner of physics. Also, it enables us to predict new physical relationships and new constants.

The key to our approach is the Absolute principle—the scale invariant equilibrium frame of reference between all physical quantities and relationships. The Absolute frame of reference, we propose, is the ultimate foundation of nature, which includes within itself all physical quantities and serves as the ultimate principle out of which all physical quantities and relationships emerge (2)-(3). It is dimensionless (formless, timeless, massless, spaceless, etc.). Mathematically, the Absolute principle can be represented as «1», or on a logarithmic scale as «zero».

Just as each number, on a mathematical scale of numbers, has zero as its starting frame of reference, so also each physical quantity, on the physics scale of quantities, has the Absolute as its starting frame of reference. The physical quantities, we propose, are natural information units of measurement of the same essence—the same reality in different containers (quanta). Like chemical elements that can be arranged on a periodic table, these quantized information constituents of nature can be arranged in order of size on a logarithmic scale as demonstrated by the physics spectrum of quantities (Figure 1). The physics spectrum, through the logarithmic slide rule of physical relationships (LSPR), can then be used to study time, space, energy, and other quantities from various vantage points, from the smallest scale of quantum electrodynamics to the largest scale of cosmic voids, and to peer forward into the future for the derivation of new constants and laws of nature.

Acknowledgment. I am deeply indebted to Professor W. Fushchych for many valuable and enlightening discussions.

Definition of Quantities

A Cross-sectional area in meter^2/radian^2

B Magnetic flux density (magnetic induction) in weber-radian^2/meter^2 or tesla

C Capacitance in coulomb/volt or farad

c Charge density in coulomb-radian^3/meter^3

D Electric flux (surface charge) density in coulombs-radian^2/meter^2

d Mass (flux) density in kilogram-radian^3/meter^3

E Electric field strength in volts-radian/meter

F (Electrical) force in (kilogram-meter-radian/second^2) or newton

f Frequency (linear) (resonance frequency) in cycles/second or hertz

G Conductance in ampere/volt or siemen

g Electric conductivity in mhos-radian/meter

H Magnetic field strength (intensity) in ampere-radian/meter

h Planck's universal constant of action in joule-second/cycle

I Moment of (rotational) inertia in kilograms-meter^2/radian^2

i Magnetic potential (electric current) in (coulombs-radian/second) or ampere

J Current (magnetic potential) density in amperes-radian^2/meter^2

L Coefficient of self-inductance in (volt-second/amp-radian) or henry

M Mass flux density in kilograms-radian^2/meter^2

m Mass (flux) in kilogram

P Electric power in joule-radian/second or watt

P_r Pressure (stress, energy density) in kilogram-radian^3/meter-second^2 or pascal

Q Volume flow rate in meter^3/second-radian^2

Q_m Mass flow rate in kilogram-radian/second

q Electric flux (charge) in ampere-second/radian or coulomb

R Resistance (capacitive or inductive reactance) in volt/ampere or ohm

R_∞ Rydberg constant in cycles^3/meter-radian^2

r Reluctance in ampere-turn/weber

S Length (radius of gyration, particle radius) in meter/radian

S_p Intensity of the wave (Pointing vector, illuminance, energy flux density) in watt-radian^2/meter^2

T Period of the harmonic motion in seconds/cycle

t Time (angular) (time constant) in seconds/radian

U Electromagnetic moment in ampers-meter^2/radian^2

V Electric potential (emf) in joules/coulomb or volt

v Velocity (linear) of motion in meters/second

V_0 Volume in meter^3/radian^3

V_g Mass (gravitational) potential, linear stopping power in joules/kilogram

W Energy (moment of force, work) in kilogram-meter^2/second^2 or joule

Y Voltage (electric potential) density in volt-radian^2/meter^2

y Mass field strength (acceleration) in meter-radian/second^2

Z Constant of universal attraction (gravitational constant) in m^3/kg-rad^2-sec^2

α Angular acceleration (mass potential density) in radian-second^2

α_f Fine structure constant in cycles/radian

ε (Electric) permittivity (of free space) in farads-radian/meter

η Coefficient of viscosity in pascal-second/radian or poise

θ Angle of rotation in radian/cycle

Ω Circular wave number in radian/meter

λ Wavelength (arc length, circumference) in meters/cycle (wave)

ω Angular velocity in radian-second

μ (Magnetic) permeability (of free space) in henry-radian/meter

1 One (Absolute frame of reference) dimensionless

Note: The unit radian (rad), for plane angle, has historically been designated as a supplementary unit. In 1980, the International Committee for Weights and Measures determined that angle and solid angle are to be regarded as dimensionless derived quantities. According to the committee, the unit radian and steradian are equivalent to the number one (1) and may be omitted in the expression for derived units. For completeness of presentation, we thought, because the angle of rotation (expressed in radian per cycle) is a physical quantity, which like other physical quantities, enters into physical relationships, it should be included here. Hence, all symbols for physical quantities, where applicable, include the radian and cycle terms.

1. *Bedrij O., Fushchych W.* On the electromagnetic structure of elementary particles masess // Dopovidi Ukrainian Academy of Sciences. Ser. A.—1991.—N 2.—P. 38-40.
2. *Bedrij O.* Fundamental constants in quantum electrodynamics // Dopovidi Ukrainian Academy of Sciences. Ser. A.—1993.—N 3.—P. 40-45.

Institute of Mathematics,
Academy of Sciences of Ukraine, Kiev

Submitted 20.10.92

МАСШТАБНАЯ ИНВАРИАНТНОСТЬ, ПРИНЦИП УНИФИКАЦИИ, ПОРЯДОК И ПОСЛЕДОВАТЕЛЬНОСТЬ ФИЗИЧЕСКИХ ВЕЛИЧИН И ФУНДАМЕНТАЛЬНЫЕ КОНСТАНТЫ

На основании принципа масштабной инвариантности физических со́отношений (уравнений, формул) предложен феноменологический подход для введения понятия «порядок» между различными физическими величинами. Установлены новые соотношения между различными физическими величинами. В рамках предложенного метода вычислены фундаментальные константы в квантовой электродинамике.

МАСШТАБНА ІНВАРІАНТНІСТЬ, ПРИНЦИП УНІФІКАЦІЇ, ПОРЯДОК І ПОСЛІДОВНІСТЬ ФІЗИЧНИХ ВЕЛИЧИН І ФУНДАМЕНТАЛЬНІ КОНСТАНТИ

На основі принципу масштабної інваріантності фізичних співвідношень (рівнянь, формул) запропонований феноменологічний підхід для введення поняття «порядок» між різними фізичними величинами. Встановлені нові співвідношення між різними фізичними величинами. В рамках запропонованого методу обчислені фундаментальні константи в квантовій електродинаміці.

APPENDIX 3

A Sample Timeline of Our Ambassadors to the '1' Miracle Zone, Giants of the Spirit Who Realized That We Are One Living Being

Name	Date	Born In
0–10		
Dipankara*		India
Zoroaster	1750–1500 Bc	Iran
Moses**	1526-1406/1350–1310Bc	Egypt
Parshvanatha	877–777 Bc	India
Isaiah	740–701 Bc	Israel
Lao Tzu	604 Bc	China
Mahavira	599–527 Bc	India
Shakyamuni Buddha	563–483 Bc	Nepal
Ananda		Nepal
Sariputra		Nepal
11–20		
Rahula	534–482 Bc	Nepal
Maha Maudgalyayana		Nepal
Subhuti		India
Maha Katyayana		India
Maha Kasyapa		India
Upali		India
Khema		India
Uppalavanna		India
Socrates	469–399 Bc	Greece

Chuang Tzu	369–286 Bc	China
21–30		
Patanjali	150 Bc	India
Philo	20 Bc–Ad 50	Egypt
Mary, Mother Of Jesus	16 Bc–	Israel
Jesus Christ	4 Bc–Ad 30	Israel
Mary Magdalene	–63	Israel
Apostle Paul	5–67	Turkey
John The Apostle	6–Ad 100	Israel
Priscilla, Saint		Italy
Aquila		Greece
Thecla, Saint	30–	Turkey
31–40		
Polycarp	69–155	Greece
Irenaeus, Saint	Second Century–202	France
Clement Of Alexandria	150–215	Egypt
Nagarjuna	150–250	India
Origen	185–254	Egypt
Plotinus	205–270	Egypt
Anthony The Great, Saint	251–356	Egypt
Porphyry	234–304	Lebanon
Abba Cronius	285–386	Egypt
Ammonas, Saint		Egypt
41–50		
Abba Poemen		Egypt
The Great, Saint		
Athanasius Of Alexandria	293–373	Egypt
Macarius The Great, Saint	295–392	Egypt
Gregory Nazianzus, Saint	329–390	Turkey
Basil Of Caesarea, Saint	330–379	Turkey
Gregory Of Nyassa, Saint	335–394	Turkey
John Chrysostom	345–407	Turkey
Jerome, Saint	347–420	Croatia

Augustine Of Hippo	354–430	Algeria
John Cassian, Saint	360–345	Bulgaria

51–60

Hilarion, Saint	–472	Egypt
Proclus	412–485	Greece
Dionysius The Areopagite	475–525	Egypt
Benedict Of Nursia, Saint	480–547	Italy
Bodhidharma	Fifth-Sixth Century	India
Gregory The Great, Saint	540–604	Italy
Muhammad Ibn Abdullah	570–632	Arabia
Dayi Daoxin	580–651	China
Zhiyan	602–668	China
Won-Hyo	617–686	Korea

61–70

Uisang	625–702	Korea
Kim Kiaokak	630–729	Korea
Beomnang	632–746	Korea
Hui-Neng (Wei-Lang)	638–713	China
Fa Tsang	643–712	China
Puji	651–739	China
Qingyuan Xingsi	660–740	China
Nan-Yang Hui-Chung	675–775	China
Nanyue Huairang	677–744	China
Yongjia Xuanjue		China

71–80

Heze Shenhui		China
Sinhaeng	704–779	Korea
Mazu Daoyi	709–788	China
Padmasambhava		Pakistan
Rabi'a Al-Basri	717–801	Iraq
P'an-Shan Pao-Chi	720–814	China
Baizhang Huaihai	720–840	China
Magu Baozhe	720–	Korea

Doui	−825	Korea
Zhizang	735–814	China

81–90

Nanquan Puyuan	748–835	China
Zhangjing Huaihui	748–835	Korea
Bai-Zhang Huai-Hai	749–814	China
Yeshe Tsogyel	757–817	Tibet
Chao-Chou Ts'ung-Shen	778–897	China
Guifeng Zongmi	780–841	China
Hyejeol	785–861	Korea
Weongam	787–869	Korea
Adi Shankara	788–820	India
Huang-Po Hsi-Yun	−850	China

91–100

Dhul-Nun Al-Misri	−861	Iran
Linji Yixuan	−866	China
Doyun	797–868	Korea
Muyeom	800–888	Korea
Chejing	804–890	Korea
Bayazid Bastami	804–874	Iran
Tung-Shan	807–869	China
Beom'il	810–889	Korea
Sahl Al-Tustari	818–896	Iran
Chiseon Doheon	824–882	Korea

101–110

Ganto Zenkatsu	828–887	China
Junayd Baghdadi	830–910	Iraq
Mansur Al-Hallaj	858–922	Iran
Leom	869–936	Korea
Gyunyeo	923–973	Korea
Symeon The New Theologian	949–1022	Turkey
Romuald, Saint	951–1025	Italy
Abu-Sa'Id Abul-Khair	967–1049	Iran

Khyungpo Naljor	978–1129	Tibet
Chang Po-Tuan	984–1082	China

111–120

Tilopa	989–1069	Bangladesh
Peter Damian, Saint	1007–1072	Italy
Theodosius Of The Caves, St.	–1074	Ukraine
Maitripa	1007–1078	Tibet
Marpa	1012–1097	Tibet
Naropa	1016–1100	India
Vajradhara Niguma	1025–	Kashmir
Bruno, Saint	1030–1101	Italy
Anselm, Saint	1033–1109	Italy
Milarepa	1052–1135	Tibet

121–130

Virupa		India
Daegak Guksa	1055–1101	Korea
Al-Ghazali	1058–1111	Iran
Sukhasiddhi		India
Gampopa	1079–1153	Tibet
Chen-Hsieh Ch'ing-Liao	1089–1151	China
Bernard Of Clairvaux	1090–1153	France
Hu-Kuo Ching-Yuan	1094–1146	China
Hildegard Of Bingen, Saint	1098–1179	Germany
Richard Of Saint Victor	–1173	France

131–140

Mokchokpa Rinchen Tsondru	1110–1170	Tibet
Dusum Khyenpa	1110–1193	Tibet
Elizabeth Of Schönau	1129–1165	Germany
Chu His	1130–1200	China
Usman Harooni	1131–1222	Iran
Maimonides (Rambam)	1135–1204	Spain
Khwaja Mu'inuddin Chishti	1141–1230	Iran
Attar Of Nishapur	1142–1220	Iran

Kyergangpa Chokyi Senge	1143–1216	Tibet
Bojo Jinul	1158–1210	Korea

141–150

Ibn Arabi	1165–1240	Spain
Qutbuddin Bakhtiyar Kaki	1173–1235	Uzbekistan
Shinran	1173–1263	Japan
Baba Farid	1173–1265	Afghanistan
Rigongpa Sangye Nyenton	1175–1247	Tibet
Chin'gak Hyesim	1178–1234	Korea
Vajrasanapa		Tibet
Ibn Al-Farid	1181–1235	Egypt
Francis Of Assisi, Saint	1182–1226	Italy
Wu-Men Hui-K'ai	1183–1260	China

151–160

Dainichi Nonin	1189–	Japan
Dogen Zenji	1200–1253	Japan
Karma Pakshi	1204–1283	Tibet
Rumi	1207–1273	Tajikistan
Mechthild Of Magdeburg	1210–1277	Germany
Saint Douceline	1214–1274	France
Nichiren	1222–1282	Japan
Ramon Lull	1232–1315	Spain
Nizamuddin Auliya	1238–1325	India
Amir Khusrau	1253–1325	India

161–170

Gertrude The Great	1256–1303	Germany
Meister Eckhart	1260–1327	Germany
Dante Alighieri	1265–1321	Italy
Nasiruddin Chiragh Dehiavi	1274–1356	India
Christina Ebner	1277–1356	Germany
Myocho Shuho	1282–1338	Japan
Rangjung Dorje	1284–1339	Tibet
Margaret Ebner	1291–1351	Germany

John Ruysbroeck	1293–1391	Belgium
Gregory Palamas, Saint	1296–1359	Greece

171–180

Gyeonghan Baeg'un	1298–1374	Korea
John Tauler	1300–1361	Germany
Henry Susso	1300–1361	Germany
Taego Bou	1301–1382	Korea
Shah Naqshband	1317–	Iran
Naong Hyegeun	1320–1376	Korea
Hafiz	1325–1390	Iran
Bassui Zenji	1327–1387	Japan
Rolpe Dorje	1340–1383	Tibet
Catherine Of Siena, Saint	1347–1380	Italy

181–190

Thang Tong Gyalpo	1361–1485	Tibet
Tsongkhapa	1367–1419	Tibet
Gihwa	1376–1433	Korea
Thomas A Kempis	1380–1471	Germany
Deshin Shekpa	1384–1414	Tibet
Ikkyu Sojun	1394–1481	Japan
Abdurrahman Jami	1414–1492	Afghanistan
Thongwa Donden	1416–1453	Tibet
Kabir	1440–1518	India
Julian Of Norwich	1342–1416	England

191–200

Catherine Of Genoa, Saint	1447–1510	Italy
Chodrak Gyatso	1454–1506	Tibet
Mira Bai	1498–1547	India
Mikyo Dorje	1507–1554	Tibet
Teresa Of Avila, Saint	1515–1582	Spain
Seosan Hyujeong	1520–1604	Korea
Yujeong	1544–1610	Korea

Taeneung	1562–1649	Korea
Trailanga Swami	1529–1687	India
Chu-Hung	1535–1615	China

201–210

John Of The Cross, Saint	1542–1591	Spain
Wangchuk Dorje	1556–1603	Tibet
Francis Bacon	1561–1626	England
Josaphat Kuntsevych	1580–1623	Ukraine
Choying Dorge	1604–1674	Tibet
Blaise Pascal	1623–1662	France
Baruch Spinoza	1632–1677	Nederland
Emanuel Swedenborg	1668–1772	Sweden
Bulleh Shah	1680–1758	Pakistan
J. S. Bach	1685–1750	Germany

211–220

Hakuin Ekaku	1685–1768	Japan
Changchub Dorge	1703–1732	Tibet
Hryhorij S. Skovorodá	1722–1794	Ukraine
Dudul Dorje	1733–1797	Tibet
W. A. Mozart	1756–1791	Austria
William Blake	1757–1827	England
Thekchok Dorje	1798–1868	Tibet
Honore De Balzac	1799–1850	France
Ralph Waldo Emerson	1803–1882	United States
Hazrat Babajan	1806–1931	Pakistan

221–230

Elizabeth B. Browning	1806–1861	England
William Henry Channing	1810–1884	United States
Walt Whitman	1819–1892	United States
Fyodor Dostoyevsky	1821–1881	Russia
Ramalinga, Swami	1823–1874	India
Lahiri Mahasaya	1828–1895	India
Bhairavi Brahmani (Yogeshwari)		India

Lyev N. Tolstoy	1828–1910	Russia
Ramakrishna Paramahamsa	1836–1886	India
Sai Baba Of Shirdi	1835–1918	India

231–240

Pranabananda Giri, Swami	–1918	India
Zigmud Gorazdowski, Saint	1845–1920	Ukraine
Annie Besant	1847–1933	England
Vladimir Soloviev	1853–1900	Russia
Sarada Devi	1853–1920	India
Yukteshwar Giri	1855–1936	India
Max Planck	1858–1947	Germany
Jozef Bilczewski	1860–1923	Poland
Rabindranath Tagore	1861–1901	India
Hazrat Tajuddin Baba	1861–1925	India

241–250

Rudolf Steiner	1861–1925	Austria
Mahavatar Babaji	1861–1935	India
Swami Vivekananda	1863–1902	India
Abraham I. Kook	1865–1935	Latvia
Andrij Sheptytsky, Saint	1865–1944	Ukraine
George Gurdjieff	1866–1949	Russia
Hryhorij Khomyshyn, Saint	1867–1947	Ukraine
Josaphat À Hordashevska, St	1869–1919	Ukraine
Mahatma Gandhi	1869–1948	India
Clementij Sheptytskyj, Saint	1869–1951	Ukraine

251–260

Upasni Maharaj	1870–1941	India
Dalun Sogaku Harada	1870–1961	Japan
D. T. Suzuki	1870–1966	Japan
Khakyab Dorje	1871–1922	Tibet
Walter Russell	1871–1963	United States
Sri Aurobindo	1872–1950	India
Therese Of Lisieux	1873–1897	France

Lalaji Maharaj	1873–1931	India
Evelyn Underhill	1875–1941	England
Chachaji Maharaj	1875–1947	India

261–270

Curuppumullage Jinarajadasa	1875–1953	Sri Lanka
Carl Gustav Jung	1875–1961	Switzerland
Albert Schweitzer	1875–1965	France
Nicholas Konrad, Saint	1876–1941	Ukraine
Josaphat Kotsylovskyj, Saint	1876–1947	Ukraine
Edgar Cayce	1877–1945	United States
Sir James Jeans	1877–1946	England
Mykyta Budka, Saint	1877–1949	Ukraine
Martin Buber	1878–1965	Austria
Mirra Alfassa (Richard)	1878–1973	France

271–280

Leonid Feodorov	1879–1935	Russia
Ramana Maharshi	1879–1950	India
Albert Einstein	1879–1955	Germany
Helen Keller	1880–1968	United States
Teilhard De Chardin	1881–1955	France
Hazrat Inayat Khan	1882–1927	India
Arthur Eddington	1882–1944	England
Khalil Gibran	1883–1931	Lebanon
Gregory Lakota, Saint	1883–1950	Ukraine
Nikos Kazantzakis	1883–1957	Greece

281–290

Omelian Kovch, Saint	1884–1944	Ukraine
Nicholas Charnetsky, Saint	1884–1959	Ukraine
Swami Ramdas	1884–1963	India
Narayan Maharay	1885–1945	India
Niels H. Bohr	1885–1962	Denmark

Emmet Fox	1886–1951	Ireland
Geoffrey Hodson	1886–1983	New Zealand
Andrew Ishchak, Saint	1887–1941	Ukraine
Erwin Schrödinger	1887–1961	Austria
Sisir Kumar Maitra	1887–1963	India

291–300

Sivananda Saraswati	1887–1963	India
Severian S. Baranyk, Saint	1889–1941	Ukraine
Peter Verhun, Saint	1890–1957	Ukraine
Maria Skobtsova, Saint	1891–1945	Latvia
Bhimrao Ramji Ambedk	1891–1956	India
Yosyf Slipij, Saint	1892–1984	Ukraine
Paramahansa Yogananda	1893–1952	India
Symeon Lukach, Saint	1893–1964	Ukraine
Aldous Huxley	1894–1963	England
Meher Baba	1894–1968	India

301–310

Philippe Barbier Saint-Hilaire	1894–1969	France
Tauji Maharaj	1894–1970	India
Oliver Reiser	1895–1974	United States
Jiddu Krishnamurti	1895–1986	India
Anna Freud	1895–1982	Austria
Yakym Senkivsky, Saint	1896–1941	Ukraine
Nicholas Tsehelsky, Saint	1896–1951	Ukraine
Ivan Sleziuk, Saint	1896–1973	Ukraine
Anandamayi Ma	1896–1982	India
Bhagavan Nityananda	1897–1961	India

311–320

Nisargadatta Maharaj, Sri	1897–1981	India
Thakur Saheb	1898–1971	India
I. K. Taimni	1898–1978	India
Nolini Kanta Gupta	1889–1983	India
Paul Brunton	1898–1981	England

Ivan Ziatyk, Saint	1899–1952	Ukraine
Maxwell Maltz	1899–1975	United States
Babuju Maharaj	1891–1987	India
Zenon Kovalyk, Saint	1903–1941	Ukraine
Olympia Bida, Saint	1903–1952	Ukraine

321–330

Gopi Krishna	1903–1984	India
Mataji Krishnabai	1903–	India
Vasyl Velychkovskyi, Saint	1903–1973	Ukraine
Indra Sen	1903–1994	Pakistan
Mahasi Sayadaw	1904–1982	Burma
Joseph Campbell	1904–1987	United States
Dag Hammarskjöld	1905–1961	Sweden
Kalu Rinpoche	1905–1989	Tibet
Volodymyr Pryjma, Saint	1906–1941	Ukraine
Bede Griffith	1906–1993	England

331–340

Marcus Bach	1906–1995	United States
Vitaliy Bayrak, Saint	1907–1946	Ukraine
Kyozan Joshu Sasaki	1907–	Japan
Douglas Harding	1909–2007	England
Teresa Of Calcutta	1910–1997	Albania
Papaji	1910–1997	Pakistan
Teodor Romzha, Saint	1911–1944	Ukraine
Laurentia Herasymiv, Saint	1911–1952	Ukraine
John A. Wheeler	1911–2008	United States
Oleksa Zarytsky, Saint	1912–1963	Ukraine

341–350

Swami Kripalvananda	1913–1981	India
David Hawkins	1913–2002	United States
Roman Lysko, Saint	1914–1949	Ukraine
Swami Satchidananda	1914–2002	India
Yizhin Norbu	–2005	Tibet

Thomas Merton	1915–1968	United States
Ahmed Kuftaro	1815–2004	Syria
Eric Butterworth	1916–2003	Canada
Swami Chidananda	1916–2008	India
Vilayat Inayat Khan	1916–2004	England

351–360

David Bohm	1917–1992	United States
Vernon Howard	1918–1992	United States
Willis Harman	1918–1997	United States
Yogi Ramsuratkumar	1918–2001	India
Tarsykia Matskiv, Saint	1919–1944	Ukraine
John Paul Ii	1920–2005	Poland
Ken Keyes Jr.	1921–1995	United States
Annemarie Schimmel	1922–2003	Germany
Itzhak Bentov	1923–1979	Czechoslovakia
Satprem (Bernard Enginger)	1923–2007	France

361–370

Robert Muller	1923–2010	Belgium
Radha Burnier	1923–	India
Thomas Keating	1923–	United States
Mataji Nirmala Devi	1923–	India
Rangjung Rigpa Dorje	1924–1981	Tibet
Sathya Sai Baba	1926–2011	India
Thich Nhat Hanh	1926–	Vietnam
Joseph Chilton Pearce	1926–	United States
Tibor Horvath	1927–2008	Hungary
Daisaku Ikeda	1928–	Japan

371–380

Joseph Girzone	1930–	United States
Bade Bhaiyya Ji	1931–2002	India
Sri Chinmoy	1931–2007	India
Lubomyr Husar	1933–	Ukraine
Chhote Bhaiyya Ji	1933–	India

Sulak Sivaraksa	1933–	Thailand
Tenzin Gyatso	1935–	Tibet
Rudolph Khoriaty	1935–	Egypt
Harold Samuel Kushner	1935–	United States
William Hungerford	1936–	United States

381–390

Jack Lenchiner	1937–	United States
Chogyam Trungpa	1939–1987	Tibet
Fritjof Capra	1939–	Austria
John White	1939–	United States
Barbara Benjamin	1939–	United States
Bokar Rinpoche	1940–2004	Tibet
Rupert Sheldrake	1942–	England
Richard Rohr	1943–	United States
Michael Snyder	1943–	United States
Aung San Suu Kui	1945–	Burma

391–400

Chun Dam	1946–	South Korea
George Rowinski	1945–	India
Peter Russell	1946–	England
Elle Collier Re	1946–	United States
Deepak Chopra	1946–	India
Eckhart Tolle	1948–	Germany
Kenneth E. Wilber	1949–	United States
Eliza Mada Dalian	1952–	Armenia
Mata Amritanandamayi	1953–	India
Vishrant	1954–	Australia

401–410

Marilyn Wilhelm		United States
Kim Jeongsoon	1954–	South Korea
Mooji	1954–	Jamaica
Igor Zupnik	1957–	Slovak Rep.

| Mother Meera | 1960– | India |
| Zia Inayat Khan | 1971– | United States |

Notes:
* Dipankara (also Dipamkara) was a Buddha who attained enlightenment prior to Gautama, the most recent historical Buddha. The number of Buddhas having lived is vast, and they are frequently as a group known under the name of Thousand Buddhas.
** We show two dates for Moses. The Bible Exodus of 1446 BC (early-date Exodus) minus eighty years for Moses' birth (1526 BC) plus forty years of wilderness for Moses' death (1406 BC). Also, the archaeological evidence Exodus of 1270 BC (late-date Exodus) minus eighty years for Moses' birth (1350 BC) plus forty years of wilderness for Moses' death (1230).

APPENDIX 4

Perfections of I AM
The Way beyond Enlightenment

The utility of a clay jar depends upon
the empty space within the jar.
The effectiveness of a human being depends on nurturing
The Way within you.

We Are Children until We Become Perfect

"You are the light of the world" (Matt. 5:14) advised the Teacher of Righteousness.

Let us consciously be the light of the world that we are.
Let us be perfect as our Beloved within us is perfect.
Let us have the love of our infinite splendor within our hearts.
Let us be our best at all times.

How do we become as He is?
How do we nurture the eternal light of the Most High—
the never-ending ocean of life that we are?

Be completely and thoroughly at peace.

Give your challenges in kindness, humility, self-responsibility, and compassion wholly to Him.

"Your Father knows what you need before you ask him" (Matt. 6:8).

With total and absolute peace in your heart and your mind permits His highest good to be your highest good. Allow Him to address your challenges His way. Whatever the outcome, let His way be your way!

This is how miracles after miracles manifest out of His power!

Let us love the Lord our God with all our heart, soul, and might.
Let us love our neighbor as ourselves, for God is in everyone.
Let us bring infinite joy and peace and freedom to your world with perfections:

1.
I AM: The life of the Most High is my life.
I AM: The will of the Most High is my will.
I AM: The light of the Most High is my light.
I AM: The peace of the Most High is my peace.
I AM: The purity of the Most High is my purity.
I AM: The service of the Most High is my service.
I AM: The dedication of the Most High is my dedication.

2.
I AM: The being of the Most High is my being.
I AM: The altruism of the Most High is my altruism.
I AM: The kindness of the Most High is my kindness.
I AM: The generosity of the Most High is my generosity.
I AM: The abundance of the Most High is my abundance.
I AM: The self-reliance of the Most High is my self-reliance.
I AM: The perseverance of the Most High is my perseverance.

3.
I AM: The work of the Most High is my work.
I AM: The fidelity of the Most High is my fidelity.
I AM: The integrity of the Most High is my integrity.
I AM: The gentleness of the Most High is my gentleness.
I AM: The magnanimity of the Most High is my magnanimity.

I AM: The understanding of the Most High is my understanding.
I AM: The trustworthiness of the Most High is my trustworthiness.

4.

I AM: The rest of the Most High is my rest.
I AM: The health of the Most High is my health.
I AM: The humility of the Most High is my humility.
I AM: The goodness of the Most High is my goodness.
I AM: The brilliance of the Most High is my brilliance.
I AM: The compassion of the Most High is my compassion.
I AM: The righteousness of the Most High is my righteousness.

5.

I AM: The justice of the Most High is my justice.
I AM: The stillness of the Most High is my stillness.
I AM: The prudence of the Most High is my prudence.
I AM: The willpower of the Most High is my willpower.
I AM: The fearlessness of the Most High is my fearlessness.
I AM: The self-sacrifice of the Most High is my self-sacrifice.
I AM: The noninterference of the Most High is my noninterference.

6.

I AM: The grace of the Most High is my grace.
I AM: The patience of the Most High is my patience.
I AM: The simplicity of the Most High is my simplicity.
I AM: The omniscience of the Most High is my omniscience.
I AM: The effortlessness of the Most High is my effortlessness.
I AM: The omnipotence of the Most High is my omnipotence.
I AM: The incorruptibility of the Most High is my incorruptibility.

7.

I AM: The mercy of the Most High is my mercy.
I AM: The courage of the Most High is my courage.
I AM: The freedom of the Most High is my freedom.
I AM: The awareness of the Most High is my awareness.
I AM: The confidence of the Most High is my confidence.
I AM: The transparency of the Most High is my transparency.
I AM: The meticulousness of the Most High is my meticulousness.

8.

I AM: The beauty of the Most High is my beauty.
I AM: The healing of the Most High is my healing.
I AM: The knowing of the Most High is my knowing.
I AM: The diligence of the Most High is my diligence.
I AM: The sacredness of the Most High is my sacredness.
I AM: The nonviolence of the Most High is my nonviolence.
I AM: The self-expression of the Most High is my self-expression.

9.

I AM: The responsibility of the Most High is my responsibility.
I AM: The commitment of the Most High is my commitment.
I AM: The consistency of the Most High is my consistency.
I AM: The equanimity of the Most High is my equanimity.
I AM: The forgiveness of the Most High is my forgiveness.
I AM: The innocence of the Most High is my innocence.
I AM: The foresight of the Most High is my foresight.

10.

I AM: The majesty of the Most High is my majesty.
I AM: The sincerity of the Most High is my sincerity.
I AM: The openness of the Most High is my openness.
I AM: The stateliness of the Most High is my stateliness.
I AM: The egolessness of the Most High is my egolessness.
I AM: The celebration of the Most High is my celebration.
I AM: The magnificence of the Most High is my magnificence.

11.

I AM: The holiness of the Most High is my holiness.
I AM: The wisdom of the Most High is my wisdom.
I AM: The caring of the Most High is my caring.
I AM: The unity of the Most High is my unity.
I AM: The truth of the Most High is my truth.
I AM: The love of the Most High is my love.
I AM: The joy of the Most High is my joy.

So be it.
And now,
Let me bow my head in peace, compassion, and gratitude,
To the Eternal Light in you,
And to the Absolute One in all.

APPENDIX 5

The Great Invocation

From the point of Light within the Mind of God
Let light stream forth into the human minds.
Let Light descend on Earth.

From the point of Love within the Heart of God
Let love stream forth into the human hearts.
May the Coming One return to Earth.

From the centre where the Will of God is known
Let purpose guide the little human wills –
The purpose which the Masters know and serve.

From the centre which we call the humanrace
Let the Plan of love and Light work out
And may it seal the door where evil dwells.

The Great Invocation is a World Prayer, an invocation for light and love, and used globally as an act of service to humanity to aid the Plan of God to all expressions on Earth. The original version released in1945 by Alice Bailey and The Tibetan, Djwhal Khul, and the newer adapted to accommodate the world's changing consciousness.

NOTES

Introduction
Your Peace, Joy, and the Miracle-after-Miracle Life

1   Saint Athanasius: www.azquotes.com/author/26896-Athanasius_
     of_Alexandria. Saint Athanasius (296–373) stated that in his book
     On the Incarnation, written about 318. He was referring to the
     course of spiritual growth, development, and *transformation*—
     divinization, or theosis, which literally means "to become divine."
     God became man. Jesus took flesh and became a human being so
     that man could through a process of catharsis (purification of mind
     and body) and theoria (illumination of consciousness) realize his/
     her *unity* with the Eternal. The transformation process comes about
     like the ocean water (the Eternal, Godhead, Absolute; see appendix
     2) becoming a cloud, rain, snow, brook, stream, river, and then part
     of the ocean water again (apotheosis, making divine, union with
     God; also see appendix 1).

2   Meister Eckhart: *The Catholic Encyclopedia, Volume V,* published
     1909. New York: Robert Appleton Company. Nihil Obstat, 1
     May 1909. Remy Laford, Censor. Imprimatur. John M. Farley,
     Archbishop of New York. Also, *Meister Eckhart's Sermons,* translated
     into English by Claud Field at Christian Classics Ethereal Library.

3   Exodus 3:14, commentary and verse meaning, "I AM WHO I . . .
     " https://www.biblehub.com/commentaries/exodus/3-14.htm.

4   St. Basil the Great Letters: *St. Basil the Great Resources Online and
     in Print,* Translated by Bl. Jackson. Also, *Works of Basil of Caesarea*
     at LibriVox (public domain audiobooks). And St. Basil the Great,

Greater Asketikon, *The Catholic Encyclopedia,* Vol. 2. New York: Robert Appleton Company, 1907.

5   Christmas Meditations—St. Basil the Great. https://www. dhsparish.com/medxmas_basil_html.

6   List of some books that did not make it into the New Testament:

Noncanonical Gospels
The Coptic Gospel of Thomas
The Gospel of the Nazareans
The Gospel of Ebionites
The Gospel According to the Hebrews
The Gospel According to the Egyptians
The Unknown Gospel (Papyrus Egerton)
The Gospel of Peter
The Gospel of Mary
The Gospel of Philip
The Gospel of Truth
The Gospel of the Savior
The Infancy Gospel of Thomas
The Proto-Gospel of James
The Epistle of the Apostles
The Coptic Apocalypse of Peter
The Second Treatese of the Great Seth
The Secret Gospel of Mark

Noncanonical Acts of the Apostles
The Acts of John
The Acts of Paul
The Acts of Thecla
The Acts of Thomas
The Acts of Peter

Noncanonical Epistles and Related Writings
The Third letter to the Corithiansdach
Correspondence of Paul and Seneca
Paul's Letter to the Laodiceans
The Letter of 1 Clement

The Letter of 2 Clement
The Letter of Peter to James
The Homilies of Clement
Treatise on the Resurrection
The Didache
The Letter of Barnabas
The Preaching of Peter
Pseudo-Titus

Noncanonical Apocalypses and Revelatory Treatises
The Sheperd of Hermas
The Apocalypse of Peter
The Apocalypse of Paul
The Secret Book of John
On the Origin of the World
The First Thought in Three Forms
The Hymm of the Pearl

7   *Love Is a Fire: The Sufi's Mystical Journey Home.* https://goldensufi. org/publications/books/love-is-a-fire.

8   Brahma—God of creation. https://www.hindudharmaforums. com/showthreads.php?2140-Brahma-God-of-creation.

9   Quote by Miyamoto Musashi: "There is nothing outside of . . . ." https://www.goodreads.com/quotes/413935-there-is . . .

10   Jesus Is Entirely and Is Everywhere (saying seventy-seven). https:// www.wisdomlib.org/christianity/compilation/ . . .

11   Meister Eckhart: *Meister Eckhart's Sermons*, translated into English by Claud Field, at Christian Classics Ethereal Library.

12   In the encounter of the burning bush (Exodus 3:14), Moses was told, in the Hebrew language that "I Am that I Am" (other translations, "I am who I am," I am what I am," or "I will be what I will be") has sent Moses to the Israelites. Also, when Philip wanted a plain sight of God, Jesus repeated the known YHWH *I am who I am* meaning: "He that hath seen me hath seen the Father" (Jn 14:9).
     In the Law of Moses, also called the Mosaic Law, for the most part, refers to the Torah or the first five books of the Hebrew Bible. The

forbidden fruit from the tree of the knowledge of good and evil (Gen 2:16–17) was off limit to mankind. In the New Testament (Rev 2:7), Jesus makes it accessible: "To those who have won the victory I will give the right to eat the fruit of the tree of life that grows in the Garden of God." Also, (in Rev 2:17) He says, "To those who have won the victory I will give some of the hidden manna."

[13] "The kingdom of God is within you" (Lk 17:21). Also, Judaism teaches that each and every human being is responsible to create the messianic age: A better world of universal peace and brotherhood on earth without war, crime, and poverty.

[14] Abraham Lincoln: www.abrahamlincolnonline.org/lincoln/speeches/1832.htm. This is Lincoln's first political declaration, which was in print in the *Sangamo Journal*. Lincoln was seeking his first seat in the Illinois General Assembly, which he lost. He was twenty-three years old and a greenhorn to the state and the public.

[15] Abraham Lincoln: https://www.azquotes.com/quotes/1404705.

[16] Renee Fleming, "God Bless America," 2020. https://www.bing.com/search?q=God+bless+america.

Where Are We Now?
God to Man and Man to God

You Have Slept for Millions and Millions of Years

[1] Kabir: Friend, Wake Up! Why Do You Go On Sleeping? https://www.poemhunter.com/poem/friend-wake-up-why . . .

The Story of the Hidden Treasure
Awakening To a New Level of Reality

[1] End Your story. Begin Your Life…: Mastering the Practice of Freedom. https://www.amazon.com/End-Your-Story-Begin-Life/dp/0615324231.

2   Gandhi: A Spiritual Biography. Walmart.com. https://www. walmart.com/ip/Gandhi-A-Spiritual . . .

Universes Come and Go
You Will Always Be in Eternity

1   Orest Bedrij, "Scale Invariance, Unifying Principle, Order and Sequence of Physical Quantities and Fundamental Constants." Ukrainian Mathematical Journal, tr. of the Proc. Inst. Math. Natl. Acad. Sci. of Ukraine (New York: Allerton Press, 1994), pp. 67–74; also, *Dopovidi: Proc. of the Nat. Acad. of Sci. of Ukr., no 4 (April 1993, pp. 67–74).*

2   As the American Civil War drew to its close amid fierce conflict, Abraham Lincoln stated this: https://azquotes.com/quote/700943.

3   Origen quote: "What man of sense will agree with the statement . . . " https://quotes.thefamouspeople.com/origen-5039.php.

4   Origen and the Holy Scriptures: copticchurch.net/topics/patrology/schoolofalex2/chapter03.html.

5   Hodson, Geoffrey, *The Christ Life from Nativity to Ascension.* London: Theosophical Publishing House, 1975. Also, *Hidden Wisdom In the Holy Bible.* Vol. 2. Wheaton: Theosophical Publishing House, 1971.

6   Moses Maimonides, Truth Control: https://www.truthcontrol. com/moses-maimonides.

7   *Zohar* III, 1528, (Soncino ed., vol. V, p. 211). The best edition of the book of *Zohar* is that by Christian Knorr von Rosenroth, with Jewish commentaries (Shulzbach, 1684). Also, The Torah of G-d—Kabbalah, Chassidism and Jewish Mysticism: https://www. chabad.org/.../jewish/The-Torah-of-G-d.htm.

8   *Bhagavad-Gita* in Sanskrit. Many English translations are available, among them that of Swami Prabhavananda and Christopher Isherwood, a Mentor paperback, 1954. Also, see bibliography for other translations.

9   *Meister Eckhart's Sermons*, translated into English by Claud Field, at Christian Classics Ethereal Library (Interalia magazine).

10   "Exploring the Nature of Consciousness," Interalia Magazine. https://www.interaliamag.org/articles/exploring-the-nature-of-consciousness.

11   Quote by Max Planck: "A new scientific truth does not . . . " https://www.goodreads.com/quotes/4079-a-new . . .

God Cannot Be Known Except By Himself
God Knows Himself

1   Meditations: Advent Sunday—Gnosis. www.gnosis.org/ecclesia/lect001.htm.

2   On Thomas Aquinas ending his writing (philosophy), https://forums.catholic.com/t/on-thomas-aquinas-ending-his-writing/441997. Also, Robert Maynard Hutchins confirms this quote in *Great Books of the Western World, vol. 19.*

Mining the Eternal Being
A Measure of Knowledge beyond the Intellect

1   Videos of it was granted to me to perceive in one instant how all th . . . www.bing.com/videos.

2   Videos of a single hour of meditation, Saint Ignatius of Loyola. www.bing.com/videos.

3   During his visit on earth, the Lord Jesus made a number of references not only about our oneness and our divinity but also *"why he is here."*

In Matthew (18:11) Jesus said, "For the Son of Man came to save the lost," and then he continues with the parable of the Lost sheep, stating, "Your Father in heaven does not want any of these little ones to be lost" (Mt 18:10–14).

In Luke (19:10), we have a repeat, "For the Son of Man came to seek and to save the lost."

Who are the lost? In Revelation (12:7–10) we have, "Then war broke out in heaven, Michael and his angels fought the dragon, who fought back with his angels; but the dragon was defeated, and he and his angels were not allowed to stay in heaven any longer. The huge dragon was thrown out! He is that old serpent, named the Devil, or Satan, that deceived the whole world. *He was thrown down to earth, and all his angels with him* [my stress]."

In Revelation (2:4–5), we obtain, "But here is what I have against you: you do not love me now as you did at first. *Remember how far you have fallen!"* (my highlighting). The angelic have forgotten their identity.

In Luke (15:11–32), we have the parable of the prodigal son.

In Matthew (7:21) and Luke (25–27), Jesus said, "Not everyone who calls me 'Lord, Lord,' will enter into the Kingdom of heaven, but only those who do what my Father in heaven wants them to do."

4   Rumi quote: "It is as if a king had sent you to . . . " https://www. azquotes.com/quote/962045 . . .

5   *Language of Jesus.* It is commonly agreed that Jesus and his disciples, for the most part, spoke Aramaic, the common language of Judaea in the first-century AD.

The Aramaic primacy. There is an old assumption that the New Testament was originally written in Greek. New evidence suggests that it was written in Aramaic, the primary starting place and source: Jesus. The Aramaic New Testament, according to George M. Lamsa, was written before the Greek version.

Aramaic to English translation. Lamsa translation remains the best known of Aramaic to English translations of the New Testament.

6   Renee Fleming, "America the Beautiful." Harvard.com. 2015. https://www.bing.com/search?q=america+the+beautiful.

7   Videos of mustard seed, Jesus. www.bing,com/videos.

8   "The Bhagavad Gita: The Practice of Meditation." https://www. thecontemplative.org/blog/bhagavad-gita-meditation-practice. In chapter 6 of the Bhagavad Gita, Krishna describes the practice of meditation.

9   Abraham Lincoln quotes (author of the Gettysburg Address). https://www.goodreads.com/author/quotes/229.Abramam_ Lincoln.

10   William James quote: "The greatest revolution of our . . . " https:// www.azquotes.com/quotes/518254.

Sanctity and Higher Degree of Perfection
Consider this an Invitation to Yourself

1   Pierre Teilhard de Chardin, "Someday, after mastering the . . . " https://www.brainyquote.com/quotes/pierre_teilhard_de_chardin_114239.
2   John Archibald Wheeler, *At Home in the Universe* (Woodbury, NY: American Institute of Physics Press, Masters of modern physics, 1994), p. 302.
3   Steven Weinberg, *Dreams of a Final Theory* (New York: Pantheon Books, 1993), p. 236.
4   *Ibid.*, p. 242.

When the Sun Is Shining Headlights Are of No Use, Become Divine Light Fast

1   Khawaja Abdul Hamid Irfani, "The Sayings of Rumi and Iqbal," Bazm-e-Rumi, 1976.

Jesus Said, "Let Him Who Seeks, Not Cease Seeking until He Finds, and When He Finds, He Will Be Troubled, and When He Has Been Troubled, He Will Marvel, and He Will Reign Over All"

Let Him Who Seeks, Not Cease Seeking Until He Finds

1   Brihadaranyaka Upanishad. www.yogananda.com.au/upa/Brihadaranyaka01_Upanishad.html.
2   St. Augustine, "The City of God," *Great Books of the Western World,* vol. 18 (Chicago: Encyclopaedia Britannica Inc., 1952). Also, videos of "Lord, I have sought you in all the temples of ..." www.bing.com/videos.
3   Catherine of Genoa quote: "My 'me' is God nor do I . . . " https.//quotefancy.com/quote/1639943/Catherine-of ...

And When He Finds, He Will Be Troubled

1 "Gospel of Thomas Saying 3." GospelThomas.com. www.early christianwritings.com/thomas/gospelthomas3.html.

2 On Thomas Aquinas ending his writing (philosophy), https://forums.catholic.com/t/on-thomas-aquinas-ending-his-writing/441997. Also, Robert Maynard Hutchins confirms this quote in *Great Books of the Western World*, vol. 19.

3 "Father Robert Barcelos, OCD: The Wisdom of Saint John of . . . " https://thespeakroom.org/2017/12/father-robert-barcelos-saint-john-of-the-cross-6.

4 Magdeburg: "My Spiritual Awakening – The day of my spiritual awakening . . . ." www.myspiritualawakening.net.

5 Magdeburg: videos of "I am in you and you are. We cannot be closer." www.bing.com/videos.

And When He Has Been Troubled, He Will Marvel

1 Tat twan asi: Sanskrit, "Thou art that" statement is commonly reiterated in the sixth chapter of the Chandogya Upanishad (c. 600 BCE), as the teacher Uddalaka Aruni tutors his son in the nature of Brahman, the supreme reality, and the Absolute Splendor.

Some Signs of the Cosmic Sense

1 The account of the experience is quoted from the *Proceedings and Transactions of the Royal Society of Canada*, 1906 (series II, vol. 12, pp. 159–196). Also, James H. Coyne, *Richard Maurice Bucke: A Sketch*. Toronto: Henry S. Saunders, 1923. *Revised edition reprinted from the transactions of the Royal Society of Canada, 1906*, pp. 26–30.

2 Richard Maurice Bucke, *Cosmic Consciousness: A Study in the Evolution of the Human Mind*, 1905. Innes editing, facsimile, 37 MB PDF file.

[3]   William James's letter is quoted in Bucke's *Cosmic Consciousness: A Study in the Evolution of the Human Mind*. Mineola, New York: Dover Publication, 2009. ISBN 9780486471907.

The Greatest Achievement
Cosmic Consciousness

[1]   Orest Bedrij, The Greatest Achievement: Miracle after Miracle the Easy Way (Philadelphia: Xlibris, 2014).

[2]   *Ibid.*, pp. 145–158.

[3]   *Ibid.*, p. 31.

Waking the Illumination of the Self and Miracles

[1]   "The Vision of Angelino of Foligno" poem. https://www.poemhunter.com/poem/the-vision-of-angelino-of-foligno/.

Where Do We Have To Go?
Blessed Are the Pure in Heart, for They Shall See God

[1]   Videos of "Many lives have you had as insects," Guru Arjan. www.bing.com/videos/.

[2]   "I am Awake," teaching of the Buddha. https://teachingsofthebuddha.com/i_am_awake.htm/.

[3]   "We have found that where science has progressed . . . " https://www.whatshouldirednext.com/quotes/arthur . . . . Also, Werner Heisenberg, *The Physicist's Conception of Nature* (New York: Harcourt and Brace, 1955).

[4]   David Bohm, *Wholeness and the Implicate Order* (London: Ark Paperback Ltd., 1983), p. 174; second part of quote from a lecture delivered at the 1983 Mystics and Scientist conference in David Lorimer, ed. "Cosmos, Matter, Life and Consciousness." The Spirit of Science: From Experiment to Experience (New York: Continuum, 1999), p. 55.

5   David Bohm, *Wholeness and the Implicate Order* (London: Ark Paperback Ltd., 1983), p. xi.

6   Max Planck. "I regard consciousness as fundamental. I . . . " https://www.azquotes.com/1056165.

7   "What does John Wheeler mean by saying "There is no 'out . . . " https://www.quora.com/What-does-John-Wheeler-mean . . .

8   "The anthropic universe – The Science Show – ABC Radio . . . " https://www.abc.net.au/radionational/programs/...

9   Niels Bohr quote: "When searching for harmony in life one . . . " https://quotefancy.com/quote/1009182/Niels-Bohr-When-searching-for-harmony-in-life . . .

10 "Schrodinger's God," Enlightened Crowd. https://enlightenedcrowd.org/schrodingers-god. Also, Erwin Schrödinger, *What Is Life?* (Cambridge: The MacMillan Company, 1946).

11 "What is Noetic Science? . . . What are the Noetic sciences . . . " www.whispy.com/noetic-science.

12 Willis W. Harman, "Business as a Component of Global Economy," *Noetic Sciences Review,* Nov. 1971.

13 Abraham Joshua Heschell, *I Asked for Wonder: A Spiritual Anthology,* edited by Samuel H. Dresner (New York: Crossroad, 2000), p. 43.

14 https://www.azquotes.com/author/6636-Abraham_Joshua_Heschel.

15 "Jiddu Krishnamurti: Enlightenment Story," Enlightened People. https://enlightened-people.com/jiddu-krishnamurti-enlightenment-story.

16 John White, ed., *What Is Enlightnment?* (Los Angeles: Jeremy P. Tarcher Inc. 1984), p.xv.

17 "Gospel of Thomas Saying 49." GospelThomas.com. www.earlychristianwritings.com/thomas/gospelthomas49.html.

18 "Gospel of Thomas Saying 50." GospelThomas.com. www.earlychristianwritings.com/thomas/gospelthomas50.html.

19 "Gospel of Thomas Saying 77." GospelThomas.com. www.earlychristianwritings.com/thomas/gospelthomas77.html.

20 "Gospel of Thomas Saying 11." GospelThomas.com. www.earlychristianwritings.com/thomas/gospelthomas11.html.

21 *The Svetasvatara Upanishad*, chapter 4. Hinduwebsite.com.

22 Charlene Leslie-Chaden, *A Compendium of the Teachings of Sathya Sai Baba* (Prasanthi Nilayam: Sai Towers Publishing, 1997), p. 215.

23 Rumi's Poems, "I died to the mineral state and became a . . . " https://www.facebook.com/rumispoems/posts/387467518313344.

24 Meister Eckhart quotes: "The seed of God is in us . . . " https://www.inspirationalstories.com/quotes . . .

25 Albert Schweitzer, quoted by Richard Carlson and Benjamin Shild in *Handbook for the Soul* (Boston: Little, Brown and Company, 1995), p. 85.

26 Ibid., p. 85.

27 Swami Vivekananda quote: "We believe that every being is . . . " https://www.azquites.com/quote/825111.

28 Hafiz: www.earthharmonyhome.com/are-you-an-elephant-with-amnesia.

29 Quote by Voltare: "Wherever my travels may lead, paradire . . . " https://www.goodreads.com/quotes/582643-wherever . . .

30 Emerson. www.goodreads.com/559902-every-man-is-a-divinity-in-disguise-a-g.

31 Meher Baba. www.en.wikiquote.org/wiki/Meher_Baba.

32 Meher Baba. *The Everything and the Nothing* (Berkeley, CA, 1971), p. 78.

33 Meher Baba. www.en.wikiquote.org/wiki/Meher_Baba.

34 Meher Baba. *Sparks from Meher Baba* (Myrtle Beach, SC: Sheriar Press, 1962), pp. 9–10.

35 "Spiritual Beings on Human Journey—Remembering Our Stardust." https://www.psychologytoday.com/us/blog/inviting . . .

36 Quote by Lao Tzu: "Be content with what you have; rejoice . . . " https://www.goodreads.com/quotes/2926-be-content-with-what-you-have-rejoice-in-th . . .

37 "Let Nothing Upset You." Blue Mountain Center of Meditation. https://www.bmcm.org/inspiration/passages/let-nothing-upset-you.

38 Matthew 6:19–21.

39 Ibn Arabi quote: "It is He who is revealed in every face . . . " https://www.azquotes.com/quote/1088826.

40 "Bhagavad Gita 7.7." www.bhagavad-gita.us/bhagavad-gita-7-7.

How Do We Get There?
Be Still and Know That I Am God

1   "Higher consciousness." Wikiquote. https://en.wikiquote.org/wiki/Higher_consciousness.

2   Quotes attributed to Epictetus. ThoughtCo. https://www.thoughtco.com/quotes-from-epictetus-121142.

3   "Thomas a Kempis Quotes." One Journey. https://onejourney.net/thomas-a-kempis-quotes.

4   November 2017. Inspiration, motivation, spiritual. www.appleseeds.org/nov_17.htm.

5   Martin Buber quote: "The beating heart of the universe is . . . " https://www.azquotes.com/quote/848683.

6   "Quotes about Distract." https://www.quotemaster.org/Distract.

7   Baruch Spinoza, "The world would be happier if men had the . . . " www.brainyquote.com/quotes/baruch_spinoza_386255.

8   "Contemplation." Friend of Silence. https://friendsofsilence.net/quote/tag/contemplation.

9   www.livinglifefully.com/joy2.htm.

10 *No more Secondhand God*. Fuller, R. Buckminster. Amazon. https://www.amazon.com//more-Secondhand-God . . .

11 Do All the Good You Can; In All the Ways You Can – Quote . . . https://quoteinvestigator.com/2016/09/24/all-good.

12 Chief Tecumseh, Shawnee. Indigenous peoples. https://www.indigenouspeople.net/tecumseh.htm.

13 "99 Thoughtful Quotes to Help You Really Say Thank You . . . " https://www.brightdrops.com/thank-you-quotes.

14 Related—Yoga International. https://www.yogainternational.com/article/view/The-Bhagavad-Gita-on-Love.

15 "When eating a fruit, think of the person who planted the . . . " https://www.lockeinyoursuccess.com/when-eating-a . . .

16 Gautama Buddha quote: "Believe nothing just because a so . . . " https://www.azquotes.com/quote/667147.

17 Dalai Lama: "If you want others to be happy, practice . . . " https://www.brainyquote.com/quotes/dalai_lama_1055551.

18 50 Hadith from the prophet Muhammad (SAW)—God does not . . . https://www.youtube.com/watch?v=8yaM28GDIcE.

19 Will Rogers quote: "There is nothing as easy as denouncing . . . "
    https://www.azquotes.com/quote/1056153.

20 Ramakrishna quote: "If you meditate on your ideal, you . . . "
    https:///www.quotefancy.com/quote/1377275/Ramarishna . . .

21 "Where shall the word be found, where will the word . . . " https://
    www.goodreads.com/quotes/473544-where . . .

22 Carl Rogers quote: "I prize the privilege of being alone." https://
    www.azquotes.com/quote/701135.

23 "You need solitude if you are to fulfill your promises." https://www.
    barnesandnoble.com/w/wonders-of . . .

24 "We must, like a painter, take time to stand back from the work."
    https://www.barnesandnoble.com/w/wonders-of . . .

25 "In solitude one can achieve a good relationship
    with oneself." https://www.barnesandnoble.com/w/
    wonders-of-solitude-dale-salwak/1002188532.

26 "If a woman is to know herself, then periods of solitude
    should . . . " https://www.barnesandnoble.com/w/
    wonders-of-solitude-dale-salwak/1002188532.

27 "Only in the oasis of silence can we drink deeply from
    our inner cup." https://www.barnesandnoble.com/w/
    wonders-of-solitude-dale-salwak/1002188532.

28 "And at some time in your life trying to be good may
    be to stop . . . " https://www.barnesandnoble.com/w/
    wonders-of-solitude-dale-salwak/1002188532.

29 "The insight we gain from solitude has very little to do
    with the . . . " https://www.barnesandnoble.com/w/
    wonders-of-solitude-dale-salwak/1002188532.

30 "Being alone gives us the space to listen again to our
    inner rhythms." https://www.barnesandnoble.com/w/
    wonders-of-solitude-dale-salwak/1002188532.

31 "Solitude is simply spending time connecting with
    ourselves." https://www.barnesandnoble.com/w/
    wonders-of-solitude-dale-salwak/1002188532.

32 "Learn to get in touch with silence within yourself and
    know that . . . " https://www.barnesandnoble.com/w/
    wonders-of-solitude-dale-salwak/1002188532.

33 "The more faithfully you listen to the voice within you, the better . . . " https://www.barnesandnoble.com/w/wonders-of-solitude-dale-salwak/1002188532.

34 "O Solitude, the soul's best friend, that man acquainted with himself . . . " https://www.barnesandnoble.com/w/wonders-of-solitude-dale-salwak/1002188532.

35 "Solitude is the furnace of transformation. Without solitude we . . . " https://www.barnesandnoble.com/w/wonders-of-solitude-dale-salwak/1002188532.

36 "You think that I am impoverishing myself by withdrawing from . . . " https://www.barnesandnoble.com/w/wonders-of-solitude-dale-salwak/1002188532.

37 "It is in deep solitude that I find the gentleness with which I can . . . " https://www.barnesandnoble.com/w/wonders-of-solitude-dale-salwak/1002188532.

38 "Devote six years to your work but in the seventh go into solitude . . . " https://www.barnesandnoble.com/w/wonders-of-solitude-dale-salwak/1002188532.

39 "Silence is precious, for it is of God. In silence all God's acts are . . . " https://www.barnesandnoble.com/w/wonders-of-solitude-dale-salwak/1002188532.

40 "In the attitude of silence the soul finds the path in a clearer light." https://www.barnesandnoble.com/w/wonders-of-solitude-dale-salwak/1002188532.

41 "We must learn to soundproof the heart against the intruding noises." https://www.barnesandnoble.com/w/wonders-of-solitude-dale-salwak/1002188532.

42 "On every mountain height is rest." https://www.barnesandnoble.com/w/wonders-of-solitude-dale-salwak/1002188532.

43 "When we are unable to find tranquility within ourselves, it is useless." https://www.barnesandnoble.com/w/wonders-of-solitude-dale-salwak/1002188532.

44 "Silence is a friend who will never betray." https://www.barnesandnoble.com/w/wonders-of-solitude-dale-salwak/1002188532.

45   "Let your mind be quiet, realizing the beauty of the world."   https://www.barnesandnoble.com/w/wonders-of-solitude-dale-salwak/1002188532.

46   "Peace of mind must come in its own time, as the waters settle."   https://www.barnesandnoble.com/w/wonders-of-solitude-dale-salwak/1002188532.

47   "Every kind of creative work demands solitude, and being alone . . . " https://www.barnesandnoble.com/w/wonders-of-solitude-dale-salwak/1002188532.

48   "Great ideas come into the world as gently as doves. Perhaps, then, if . . . " https://www.barnesandnoble.com/w/wonders-of-solitude-dale-salwak/1002188532.

49   "If you are a writer you locate yourself behind a wall of silence and . . . " https://www.barnesandnoble.com/w/wonders-of-solitude-dale-salwak/1002188532.

50   "When the mind is very quiet, completely still, when there is not a . . . " https://www.barnesandnoble.com/w/wonders-of-solitude-dale-salwak/1002188532.

51   "A public man, though he is necessarily available at many times . . . " https://www.barnesandnoble.com/w/wonders-of-solitude-dale-salwak/1002188532.

52   "The hours which I have spent alone with Mr. Edison have brought . . . " https://www.barnesandnoble.com/w/wonders-of-solitude-dale-salwak/1002188532.

53   "Without great solitude no serious work is possible."   https://www.barnesandnoble.com/w/wonders-of-solitude-dale-salwak/1002188532.

54   "I prefer to get up very early in the morning and work."   https://www.barnesandnoble.com/w/wonders-of-solitude-dale-salwak/1002188532.

55   "There is nobody else like you. The more you can quiet your own . . . " https://www.barnesandnoble.com/w/wonders-of-solitude-dale-salwak/1002188532.

56   "Obviously, if we are to experience insights from our consciousness . . . " https://www.barnesandnoble.com/w/wonders-of-solitude-dale-salwak/1002188532.

57   "Talent is nurtured in solitude; character is formed on the stormy . . . " https://www.barnesandnoble.com/w/wonders-of-solitude-dale-salwak/1002188532.

58   "Solitude is the nurse of enthusiasm enthusiasm is the true part of . . . " https://www.barnesandnoble.com/w/wonders-of-solitude-dale-salwak/1002188532.

59   "The best remedy for those who are afraid, lonely, or unhappy is to . . . " https://www.barnesandnoble.com/w/wonders-of-solitude-dale-salwak/1002188532.

60   "It all adds up to one thing: Peace, silence, solitude. The world . . . " https://www.barnesandnoble.com/w/wonders-of-solitude-dale-salwak/1002188532.

61   "When you pray, go into your room and shut the door and pray." https://www.barnesandnoble.com/w/wonders-of-solitude-dale-salwak/1002188532.

62   "When I need solitude, I turn off the phone and fax and sit until . . . " https://www.barnesandnoble.com/w/wonders-of-solitude-dale-salwak/1002188532.

63   "Within you there is a stillness and a sanctuary to which you can . . . " https://www.barnesandnoble.com/w/wonders-of-solitude-dale-salwak/1002188532.

64   "What a strange power there is in silence. How many resolutions . . . " https://www.barnesandnoble.com/w/wonders-of-solitude-dale-salwak/1002188532.

65   "The victories of speech have been many, but the victories of silence . . . " https://www.barnesandnoble.com/w/wonders-of-solitude-dale-salwak/1002188532.

66   "There is something greater and purer than mouth utters. Silence . . . " https://www.barnesandnoble.com/w/wonders-of-solitude-dale-salwak/1002188532.

67   "He that would live in peace and ease must not speak all he knows . . . " https://www.barnesandnoble.com/w/wonders-of-solitude-dale-salwak/1002188532.

68   "There are times when silence is the most sacred of responses." https://www.barnesandnoble.com/w/wonders-of-solitude-dale-salwak/1002188532.

69 "Better to remain silent and be thought a fool than to speak out . . . " https://www.barnesandnoble.com/w/wonders-of-solitude-dale-salwak/1002188532.

70 "Keep silent, because the world of silence is a vast fullness." https://www.barnesandnoble.com/w/wonders-of-solitude-dale-salwak/1002188532.

71 "Even a fool who keeps silent is considered wise . . . https://www.barnesandnoble.com/w/wonders-of-solitude-dale-salwak/1002188532.

72 "Silence alone can bring two hearts closer together." https://www.barnesandnoble.com/w/wonders-of-solitude-dale-salwak/1002188532.

73 "Talking is a loss of power." https://www.barnesandnoble.com/w/wonders-of-solitude-dale-salwak/1002188532.

74 "Silence is unceasing eloquence. It is the best language." https://www.barnesandnoble.com/w/wonders-of-solitude-dale-salwak/1002188532.

BIBLIOGRAPHY

Abbott, Walter M., and Chapman, Geoffrey, eds. *The Documents of Vatican II*, with notes by Protestant and Orthodox authorities, 1966.

Acklom, George Moreby. "The Man and the Book." In Maurice R. Bucke, *Cosmic Consciousness: A Study in the Evolution of the Human Mind*. New York: E. P. Dutton & Co., 1901 and 1975.

Adler, M. J., editor-in-chief. *The Great Ideas: A Syntopicon of Great Books of the Western World*. Chicago: Encyclopaedia Britannica Inc., 1971.

Affifi, Abu'l-'Ala. The Mystical Philosophy of Muhyid'Din Ibnul-'Arabi. Cambridge, 1936.

Afterman, Allen. *Kabbalah and Consciousness*. Riverdale, NY: Sheep Meadow Press, 1992.

Amaldi, E. "The Unity of Physics." *Physics Today*. September 1973.

Ambjorn, J. and Wolfram, S. "Properties of the Vacuum. 1. Mechanical and Thermodynamic." *Ann. Phys.* 147 (1983): 1–32.

———. "Properties of the Vacuum. 2. Electrodynamic." *Ann. Phys.* 147 (1983): 33–56. Andrews, Allan A. *The Teachings Essential for Rebirth: A Study of Genshin's Ojoyoshu*. Tokyo: Sophia University Press, 1973.

Anesaki, Masaharu. *Nichiren, the Buddhist Prophet*. Cambridge, MA: Harvard University Press, 1916.

Ansari, Muhammad Abdul-Haq. Sufism and Shari'ah: A Study of Shaykh AhmadSirhindi's Effort to Reform Sufism. Leicester, England: The Islamic Foundation, 1986.

Aquinas, Thomas. *On Being and Essence*. Translated by Armand Maurer. Toronto: The Pontifical Institute of Medieval Studies, 1949.

———. "Summa Theologica." Vol. 19, *The Great Books of the Western World*. Chicago: Encyclopaedia Britannica Inc., 1971.

Aristotle. *Topics,* vol. 8, bk. 4, ch. 1, *The Great Books of the Western World.* Chicago: Encyclopaedia Britannica Inc., 1971.

Armstrong, Karen. *A History of God.* New York: Ballantine Books, 1993.

Ashvagosha. *The Awakening of Faith.* Translated by D. T. Suzuki. Chicago: Open Court, 1900.

Assagioli, Roberto. *La Vie dello Spirito.* Rome: G. Filipponio, 1974.

Athenagoras. "A Plea for the Christians." Translated by Rev. B. P. Pratten. Vol. 2, *Fathers of the Second Century.* Peabody, MA: Hendrickson Publishers, 1999.

Atiyah, Michael. "Topology of the Vacuum." *The Philosophy of Vacuum.* Oxford: Oxford University Press, 1991.

Attar, Fariduddin. *Muslim Saints and Mystics.* Translated by A. J. Arberry. London: Routledge and Kegan Paul, 1966.

Augustine. "The City of God." Vol. 18, *Great Books of the Western World.* Chicago: Encyclopaedia Britannica Inc., 1952.

———. *On the Gospel of St. John.* Grand Rapids: Wm B. Eerdmans Publishing Co., 1987.

Aurobindo, Sri. *The Light Divine.* New York: The Sri Aurobindo Library Inc., 1949.

———. *Essays on the Gita.* Pondicherry: Sri Aurobindo Ashram Press, 1950.

———. *The Ideal of Human Unity.* New York: E. P. Dutton & Co., 1950.

———. *The Mind Light.* New York: E. P. Dutton & Co., 1953.

———. *The Life Divine.* Pondicherry: Sri Aurobindo Ashram Press, 1960.

———. *Birth Centenary Library.* 30 vols. Pondicherry, India: Sri Aurobindo Ashram Press, 1972.

———. *The Future Evolution of Man.* Wheaton, IL: Theosophical Publishing House, 1974.

Baba, Meher. *God to Man and Man to God.* Edited by C. B. Purdom. Myrtle Beach, SC: Sheriar Press, 1975.

———. *Sparks from Meher Baba.* Myrtle Beach, SC: Sheriar Press, 1962.

Baba, Satya Sai. *Teachings of Sri Satya Sai Baba.* Edited by Roy Eugene Davis. Lakemont, GA: CSA Press, 1974.

Bailey, Alice A. *The Consciousness of the Atom*. New York: Lucis Publishing Company, 1922.

———. *The Soul and Its Mechanism*. London: Lucis Trust, 1971.

———. *From Intellect to Intuition*. London: Lucis Trust, 1971.

Bandera, Cesareo. The Sacred Game: The Role of the Sacred in the Genesis of Modern Literary Fiction. Penn State Studies in Romance Literatures.

Barnstone, Willis, ed. "The Gospel of Thomas." *The Other Bible*. San Francisco: HarperCollins Publishers, 1984.

Barrett, C. K. From First Adam to Last, A Study in Pauline Theology. New York: Scribner, 1962.

Basler, Roy P. and others. *The Collected Works of Abraham Lincoln*. 8 vols. New Brunswick: Rutgers University Press, 1953.

Bedrij, Orest. *Yes, It's Love: Your Life Can Be a Miracle*. New York: Pyramid Publications, 1974.

———. *One*. San Francisco: Strawberry Hill Press, 1977.

———. *You*. Warwick, NY: Amity House, 1988.

———. "Grand Unification of the Science of Physics through the Cosmolog." *Abstracts: Amer. Assoc. for the Adv. of Sci.* Annual Meeting, 1990.

———. "Fundamental Constants in Quantum Electrodynamics." *Dopovidi: Proceedings of the National Academy of Sciences of Ukraine*, no. 3 (March 1993).

———. "Scale Invariance, Unifying Principle, Order and Sequence of Physical Quantities and Fundamental Constants." *Dopovidi: Proc. of the Nat. Acad. of Sci. of Ukr.*, no. 4 (April 1993).

———. "Connection of ϖ with the Fine Structure Constant." *Dopovidi: Proc. of the Nat. Acad. of Sci. of Ukr.*, no. 10 (1994).

———. "Revelation and Verification of Ultimate Reality and Meaning through Direct Experience and the Laws of Physics." *Ultimate Reality and Meaning*, University of Toronto Press, vol. 23, no. 1, pp. 36–84 (March 2000).

———. La Preuve Scientifique de L 'Existence de Dieu. Montreal: Courteau Louise Ed., 2000.

———. "New Relationships and Measurements for Gravity Physics." Vol. 43, pt. 2, Proc. of the Fourth Inter. Conf Symmetry in

Nonlinear Mathematical Physics, Proc. of Nat. Acad. of Sci. of Ukr. Institute of Mathematics, 2002.

———.*Seeing God Face to Face.* Philadelphia: Xlibris, 2004.

———.Celebrate Your Divinity: The Nature of God and the Theory of Everything. Philadelphia: Xlibris, 2005.

———. '**1**': The Foundation and Mathematization of Physics. Philadelphia: Xlibris, 2008.

———.Living Your Divine Life: Experience God's Glory, Absolute Happiness, and Great Prosperity. Philadelphia: Xlibris, 2009.

Bedrij, O., and Fushchych, W. I. "On the Electromagnetic Structure of Elementary Particles Masses," in Russian. *Doklady, Ukr. SSR Academy of Sciences*, no. 2 (February 1991).

———. "Fundamental Constants of Nucleon-Meson Dynamics." *Dopovidi: Proc. of the Nat. Acad. of Sci. of Ukr,* no. 5 (1993).

———. "Planck's Constant Is Not Constant in Different Quantum Phenomena." *Dopovidi: Proc. of the Nat. Acad. of Sci. of Ukr,* no. 12 (1995).

Bell, J. S. "On the Einstein-Podolsky-Rosen Paradox." *Physics* 1 (1964): 195.

———. "On the Problem of Hidden Variables in Quantum Mechanic." *Reviews of Modern Physics* 38 (1966): 447.

———. *Speakable and Unspeakable in Quantum Mechanics.* Cambridge: Cambridge University Press, 1987.

———. *Collected Papers in Quantum Mechanics.* Cambridge: Cambridge University Press, 1987.

Benardete, José. *Infinity.* Oxford: Clarendon Press, 1964.

Benjamin, Barbara. *Face to Face.* Yonkers, NY: Nepperham Press, 2009.

Besant, Annie. *The Self and Its Sheaths.* Adyar, Madras, India: Theosophical Publishing House (TPH), 1948.

———*An Autobiography.* TPH: Adyar, Chennai, 1984.

———*Esoteric Christianity.* Preface. Adyar, Chennai: TPH, 1989.

———*Thought Power.* Wheaton, IL: TPH, 1988.

Bhagavad-Gita, The in Sanskrit. Many English translations are available, among them that of Swami Prabhavananda and Christopher Isherwood, a Mentor Paperback, 1954; that of Ann Stanford, New York: Herder & Herder, 1971; that of P Lal, Calcutta: Writers Workshop, 1965; and that of Swami Nikhilanada, New York:

Ramakrishna-Vivekananda Center, 1952. Also a complete edition with original Sanskrit text by His Divine Grace A. C. Bhaktivedanta Swami Prabhupada. Bhatnagar, R. S. *Dimensions of Classical Sufi Thought.* Delhi: Motilal Banarsidass, 1984.

Bible, Holy: From the Ancient Eastern Text. George M. Lamsa's Translation from the Aramaic of the Peshitta. Philadelphia: A. J. Holman Co., a division of B. Lippincott Co., 1933.

Bible, The Holy: New International Version. Grand Rapids, MI: Zondervan Bible Publishers, 1978.

Bible, The Holy. New Revised Standard Version. London: Collins Publishers, 1989.

Bible, The New American. Translated from the Original Languages with Critical Use of All the Ancient Sources. Washington DC: Confraternity of Christine Doctrine, 1970.

Bitbol, M. Schrödinger's Philosophy of Quantum Mechanics. Kluwer, 1996.

Blake, William. *The Complete Writings of William Blake.* Edited by GeoffreyKeynes. Oxford: Oxford University Press, 1969.

Blavatsky, H. P. *The Theosophical Glossary.* Los Angeles: The Theosophy Co., 1930.

———. *Collected Writings.* Fifteen vols. Wheaton, IL; Adyar, Chennai, India: Theosophical Publishing House, 1966–91.

———. *The Secret Doctrine.* Adyar, Madras, India: Theosophical Publishing, 1987.

Boehme, Jacob. *The Incarnation of Jesus Christ.* Translated by J. R. Earle. London: Constable, 1934.

Bohm, David. *Quantum Theory.* New York: Prentice Hall, 1951.

———. *The Special Theory of Relativity.* New York: W. A. Benjamin, 1965.

———. *Wholeness and the Implicate Order.* Boston: Routledge & Kegan Paul Ltd., 1980. Reprint. London: Associated Book Publishers Ltd.; Ark Paperbacks Ltd., 1983.

Bohm, D., and Hiley, B. The Undivided Universe: An Ontological Interpretation of Quantum Theory. London: Routledge, 1993.

Bohr, N. *Atomic Theory and the Description of Nature.* Cambridge: Cambridge University Press, 1934.

————. Essays 1958–1962 on Atomic Physics and Human Knowledge. New York: Wiley-Interscience, 1963.

Bokser, Rabbi Ben Zion, ed. and trans. *The Essential Writings of Abraham Isaac Kook.* Warwick, NY: Amity House, 1988.

Bolzano, Bernard. *Paradoxes of the Infinite.* New Haven, CT: Yale University Press, 1950.

Boole, George. An Investigation of the Laws of Thought on Which Are Founded the Mathematical Theories of Logic and Probabilities. 1854.

Bromley, D. A., ed. "The Unity of Physics." *Physics in Perspective.* Washington DC: National Academy of Sciences, 1972.

Bruno, Giordano. *On the Infinite Universe and Worlds.* Translated by Dorothy Singer. New York: Greenwood Press, 1968.

Brunton, Paul. *The Hidden Teaching Beyond Yoga.* New York: Samuel Weiser, 1972.

Bruteau, Beatrice. *Worthy Is the World: The Hindu Philosophy of Sri Aurobindo.* Rutherford, NJ: Fairleigh Dickinson University Press, 1971.

————. Evolution Toward Divinity: Teilhard de Chardin and the Hindu Traditions. Wheaton, IL: Theosophical Publishing House, 1974.

Bub, Jeffrey. *Interpreting the Quantum World.* Cambridge: Cambridge University Press, 1997.

Buber, Martin, *Hasidism.* New York: Philosophical Library, 1997.

Bucke, Maurice R. *Cosmic Consciousness: A Study into Evolution of the Human Mind.* New York: E. P. Dutton & Co., 1901 and 1975.

Burghardt, Walter J. *The Image of God in Man according to Cyril of Alexandria.* Washington DC: Catholic University of America, 1957.

Buswell, Robert E. Jr., trans. and ed. *The Korean Approach to Zen: The Collected Works of Chinul.* Honolulu: University of Hawaii Press, 1983.

————. *The Zen Monastic Experience.* Princeton, NJ: Princeton University Press, 1992.

Cantor, Georg. *Contributions to the Founding of the Theory of Transfinite Numbers.* Translated by Philip E. B. Jourdain. La Salle, IL: Open Court, 1952.

Capra, Fritjof. *The Tao of Physics.* 3rd ed. Boston: Shambhala Publications Inc., 1991.

———. *The Turning Point.* New York: Simon & Schuster, 1982.

Capra, Fritjof and Steindl-Rast, David with Matus, Thomas. *Belonging to the Universe: Explorations on the Frontiers of Science and Spirituality.* San Francisco: HarperCollins Publishers, 1991.

Carey, Ken. *Vision.* Kansas City, MO: Uni Sun, 1985.

———. *Starseed: The Third Millennium.* San Francisco: HarperCollins Publishers, 1991.

Carpenter, Edward. *The Drama of Love and Death.* London: George Allen & Unwin Ltd.

Carter, Robert E. The Nothingness beyond God: An Introduction to the Philosophy of Nishida Kitaro. New York: Paragon House, 1989.

Catherine of Genoa. *Vita Mirabile e Dottnna Celeste de Santa Catherina de Genova.* Insieme Col Trattato del Purgatorio e col Dialogo Della Santa, 1743.

Catherine of Siena. *The Divine Dialogue of Saint Catherine of Siena.* Translated by Alger Thorold. 2nd ed. London, 1926.

Chaden, Charlene Leslie. *A Compendium of the Teachings of Sathya Sai Baba.* Prasanthi Nilayam: Sai Towers Publishing, 1997.

Champawat, Narayan. "Rabindranath Tagore." In *Great Thinkers of the Eastern World.* Edited by Ian P. McGreal. New York: HarperCollins Publishers, 1995.

Chan, Wing-Tsit, ed. and trans. *The Platform Scripture.* New York: St. John's University Press, 1963. An unabridged translation of the Tun-huang (Dunhuang) manuscript, found in a cave in Dunhuang, northwest China, in 1900.

———. trans. and comp. *A Source Book in Chinese Philosophy.* Princeton, NJ: Princeton University Press. 1963. Chapter 28 discusses Chou and gives a variety of selections from his books.

Chang, Garma C. C. The Buddhist Teaching of Totality: The Philosophy of Hwa Yen Buddhism. University Park, PA: Penn State Press, 1971.

Chapple, Christopher Key and Yogi Ananda Viraj (Eugene P. Kelly Jr.), trans. *The Yoga Sutras of Patanjali: An Analysis of the Sanskrit with accompanying English translation.* Delhi: Sri Satguru Publications, 1990.

———. "Mahavira." In *Great Thinkers of the Eastern World,* edited by Ian P. McGreal. HarperCollins Publishers, 1995.

Chardin, Pierre Teilhard de. *The Appearance of Man.* New York: Harper & Row, 1956.

———. *The Divine Milieu.* New York: Harper & Row, 1956.

———. *The Making of a Mind.* New York: Harper & Row, 1956.

———. *The Phenomenon of Man,* trans. by Bernard Wall. London: Collins, 1959.

———. *The Future of Man.* New York: Harper & Row, 1959.

———. *Building the Earth.* Wilkes-Barre, PA: Dimension Books, 1965.

———. *Hymn of the Universe.* New York: Harper & Row, 1965.

———. *Activation, of Energy.* New York: Harcourt Brace Jovanovich, 1971.

———. *Christianity and Evolution.* New York: Harcourt Brace Jovanovich, 1971.

———. "The Evolution of Chastity." *Toward the Future.* New York: Harcourt Brace Jovanovich, 1975.

Cheney, Sheldon. *Men Who Have Walked With God.* New York: Alfred A. Knopf, 1945.

Chittick, William C. *The Sufi Path of Knowledge.* Albany: State University of New York Press, 1989.

Chuang Tzu. *The Book of Chuang Tzu,* trans. Martin Palmer with Elizabeth Breuilly. London: Arkana, 1996.

Chuang-Tzu. Musings of a Chinese Mystic. London, 1920.

Chung, Bruya. *Zhuangzi Speaks!* Princeton, NJ: Princeton University Press, 1992.

Chu Ta-kao, trans. *Tao-Te Ching.* Boston: Mandala Books, 1982.

Cidade Calelixnese manuscript, found in Oxyrynchus, Egypt. Located in theBritish Library, Department of Manuscripts, London.

Cleary, Thomas. *The Dhammapada: The Sayings of Buddha.* New York: Bantam Books, 1994.

Clement of Alexandria. "The Stromata, or Miscellanies," *Ante-Nicene Fathers.* Peabody, MA: Hendrickson Publishers, 1999. Vol. 2., in Eusebius, *Ecclesiastical History* 6.14.7.

Confraternity of Christian Doctrine. The New American Bible. New York: Catholic Book Publishing Co., 1986.

Conze, Edward, ed. and trans. *Buddhist Wisdom Books: The Diamond Sutra and Heart Sutra*. London: George Allen & Unwin, 1958.

———. *Buddhist Scriptures*. Harmondsworth: Penguin Books Ltd., 1959. Corbin, Henri. "Imagination créatrice et prière créatrice dans le soufismed'Ibn 'Arabi." *Eranos-Jahrbuch* 25 (1956).

———. *Creative Imagination in the Sufism ofIbn 'Arabi*. Princeton: Princeton University Press, 1969.

Dalai Lama, H.H. the XIV, *The Universe in a Single Atom*. Morgan Road Books, 2005.

Danielou, Jean, and Herbert Musurrillo, From Glory to Glory: Texts from Gregory of Nyssa's Mystical Writings. London, 1961.

Das, Bhagavan, *The Essential Unity of All Religions*. Quest Book Edition. Wheaton: Theosophical Publishing House, 1973.

Datta, Dhirenda Mohan. *The Philosophy of Mahatma Gandhi*. Madison: The University of Wisconsin Press, 1972.

———. *Six Ways of Knowing*. Calcutta: University of Calcutta, 1972.

Dauben, Joseph W *Georg Cantor, His Mathematics and Philosophy of theInfinite*. Cambridge: Harvard University Press, 1979.

Davidson, H. "Avicenna's Proof of the Existence of God as a Necessarily Existent Being." In *Islamic Philosophical Theology*. Edited by P Morewedge. Albany: SUNY Press, 1979.

Davidson,R.J. and Harrington, Anne. *Visions of Compassion*. New York: Oxford University Press, 2002.

Descartes, Rene. "Rules for the Direction of the Mind," trans. by Elizabeth S. Haldane and G. R. T. Ross. Vol. 31, *The Great Books of the Western World*. Chicago: Encyclopaedia Britannica Inc., 1971.

Deussen, Paul. *The Philosophy of the Upanishads*. Edinburgh: T&T Clark, 1906.

Dhammapada. *Dhammapada: Wisdom of the Buddha,* trans. by Harischandra Kaviratna. Pasadena, CA: Theosophical University Press. 1889.

———. The Dhammapada: With Introductory Essays, Pail Text, English Translation and Notes. Translated by S. Radakrishnan. London: Oxford University Press, 1966.

Digha Nikaya. *Thus Have I Heard: The Long Discourses of the Buddha*. Translated by Maurice Walshe. London: Wisdom Publications, 1987.

Dirac, P A. M. *Quantum Mechanics.* 4th ed. Oxford: Clarendon Press, 1958.

———. *Directions in Physics.* New York: John Wiley and Sons, 1978.

Donald, David Herbert. *We Are Lincoln Men: Abraham Lincoln and HisFriends.* New York: Simon & Schuster, 2003.

Dossey, Larry, MD. *Space, Time & Medicine.* Boston: New Science Library, Shambhala, 1985.

Dowman, Keith, trans. Sky Dancer: The Secret Life and Songs of the Lady Yeshe Tsogyel. London: Routledge & Kegan Paul, 1984.

Downs, Robert B. *Books That Changed the World.* New York: The New American Library Inc., 1956.

Dundas, Paul. *The Jainas.* London: Routledge, 1992.

DuNouy, Lecomte. *Human Destiny.* New York: Longmans, Green & Co., 1947.

Dyson, Freeman J. *Infinite in All Directions.* New York: Harper & Row, 1988.

Eckhart, Meister. *Meister Eckhart: A Modern Translation,* trans. Raymond Bernard Blankney. New York: Harper Torchbook, 1941.

———. *Works,* trans. by C. B. Evans. London, 1924.

Eddington, Sir Arthur Stanley. *The Mathematical Theory of Relativity.* Cambridge, MA: Cambridge University Press, 1923.

———. The Nature of the Physical World. New York: Macmillan, 1929.

———. *Science and the Unseen World.* New York: Macmillan, 1929.

———. *Fundamental Theory.* Cambridge: Cambridge University Press, 1946.

———. *Space, Time and Gravitation.* New York: Harper Torchbooks, 1959.

Ehrman, Bart D. Lost Scriptures: Books That Did Not Make It into the NewTestament. Oxford: Oxford University Press, 2003.

———. Lost Christianities: The Battles for Scripture and the Faiths We Never Knew. Oxford: Oxford University Press, 2003.

Einstein, Albert. "Prinzipielles rur Allgemeinen Relativitaetstheorie [Principles Concerning the General Theory of Relativity]." *Ann d. Physik* 55 (1918).

———. *Über den Äther.* Schweizerische Naturforschende Gesellschaft Verhanflungen, 105 (1924): 85–93. For translation, see S. Saunders and H. R. Brown 1991, below.

———. "Religion and Science," New York, *NY Times Magazine,* November9, 1930.

German text published in *The Berliner Tageblatt,* November 11, 1930.

———. *Mein Weltbild.* Amstterdam: Querido Verlag, 1934.

———. "Education for Independent Thought." New York: *New York Times,* October 5, 1952.

———. "Generalization of Gravitation Theory." *The Meaning of Relativity.* Reprint of appendix 2 from 4th ed. Princeton: Princeton University Press, 1953.

———. Message conveyed at Leyden, Holland, 1953, for the honor of the 100th anniversary of the birth of Lorentz. Published in *Mein Weltbild.* Zurich: Europa Verlag, 1953.

———. *The New York Post,* November 18, 1972.

Einstein, A., B. Podolsky, and N. Rosen. *Phys. Rev.* 45 (1935): 777.

Eliot, T. S. *Four Quarters.* London: Faber and Faber, 1944.

———. *Collected Poems.* London: Faber and Faber, 1963.

Emerson, Ralph Waldo. *The Works of Ralph Waldo Emerson.* Roslyn, NY:Black's Readers Service, 2000.

———. *The Journals and Miscellaneous Notebooks of Ralph Waldo Emerson.* Edited by William H. Gilman, et al. 16 vols. Cambridge: Harvard University Press. 1960–1982.

Erkes, Eduard. *Ho-shangKung's Commentary on Lao-tse.* Ascona, Switzerland: Artibus Asiae, 1958.

d'Espagnat, Bernard. *Conceptual Foundations of Quantum Mechanics.* 2nd ed. Reading, MA: W. A. Benjamin, 1976.

———. "The Quantum Theory and Reality." *Scientific American.,* November 1979, 158–181.

———. *Veiled Reality.* Reading, MA: Addison-Wesley, 1995.

Evans-Wentz, W. Y., ed. *Tibet's Great Yogi Milarepa.* London: Oxford University Press, 1951.

———. *The Tibetan Book of the Great Liberation.* London: Oxford University Press, 1954.

———. *Tibetan Yoga and Secret Doctrines.* London: Oxford University Press, 1958.

Fehrenbacher, Don, ed. *Abraham Lincoln: Speeches and Writings 1832–58,* and *Abraham. Lincoln: Speeches and Writings 1859–65.* Library ofAmerica, two-volume set, as well as the one-volume edition. New York: Vintage, 1965.

Feng, English. *Chuang Tsu: The Inner Chapters.* New York: Vintage Books, 1974.

Feuerstein, Georg. *Yoga-Sutra of Patanjali: A New Translation and Commentary.* Feynman, Richard. *QED.* Princeton, NJ: Princeton University Press, 1985.

———. "The Distinction of Past and Future." In *The Character of Physical Law.* Cambridge, MA: The MIT Press, 1965.

———. "The Distinction of Past and Future," in *The World Treasury of Physics, Astronomy, and Mathematics.* Edited by Timothy Ferris. Boston: Little, Brown and Company, 1991.

Feynman, R. and Weinberg, S. *Elementary Particles and the Laws of Physics.* Cambridge: Cambridge University Press, 1999.

Finegan, Jack. *Light from the Ancient Past.* Princeton: Princeton University Press, 1946.

Finkelstein, D. "Theory of Vacuum." *The Philosophy of Vacuum.* Oxford: Oxford University Press, 1991.

Fleming, G. N. *The Vacuum on Null Planes.* Presented to the 1987 Oxford University Symposium on the Vacuum in Quantum Field Theory.

Foard, James Harlan. *Ippen and Popular Buddhism in Kamakura Japan.* PhD diss., Stanford University, 1977. Ann Arbor, MI: Xerox University Microfilms.

Fox, Emmet. The Sermon on the Mount: A General Introduction to Scientific Christianity in the Form of a Spiritual Key to Matthew V, VI, and VII. New York: Harper & Row, 1938.

Fremantle, Anne. *Woman's Way to God.* New York: St. Martin's Press, 1977.

Friedman, M., ed. Martin Buber's Life and Work: The Early Years 1878–1923.New York: E. P. Dutton, 1981.

Fung, Yu-lan, trans. Chuang Tzu, A New Selected Translation with an Exposition of the Philosophy of Kuo Hsing. Shanghai: Commercial Press, 1933.

―――. *A History of Chinese Philosophy.* Translated by Derk Bodde. 2 vols. Princeton: Princeton University Press, 1953.

―――. *A Taoist Classic: Chuang Tzu.* Beijing: Foreign Language Press, 1989. Gabor, Dennis. *Inventing the Future.* Harmondsworth, England: Penguin, 1964.

Galloway, Allan D. *The Cosmic Christ.* New York: Harper Brothers, 1951.

Gandhi, Mohandas K. *The Way to God.* (The original title of this work is*Pathway to God).* Berkeley, CA: Berkeley Hills Books, 1999.

Gell-Mann, Murray, and Ne'eman, Yuval. *The Eightfold Way.* New York: W. A. Benjamin Inc., 1964.

Gill, Pritam Singh. *The Doctrine of Guru Nanak.* Jullundher: New Book Company, 1969.

Gilson, Etienne. The Mystical Doctrine of Saint Bernard. London: 1940.

Girard, Rene. *Deceit, Desire & the Novel.* Baltimore: The John HopkinsUniversity Press, 1961.

―――. *Things Hidden Since the Foundation of the World.* Stanford, CA: Stanford University Press, 1978.

Glasberg, R. "Internal and External Perspectives on Immediate and Ultimate Reality: Toward the Unity of Knowledge." *Ultimate Reality and Meaning* 22 (1999): 2-–.

Gödel Kurt. *The Consistency of the Continuum Hypothesis.* Princeton, NJ: Princeton University Press, 1940.

―――. *Collected Works.* Vols. I (II) Feferman, Solomon, et al., eds. New York: Oxford University Press, 1986 (1990).

―――. On Formally Undecidable Propositions of Prtincipia Mathematica and RelatedSystems. Translated by B. Meltzer. New York: Dover, 1992.

Goddard, Dwight, ed. *A Buddhist Bible.* Boston: Beacon Press, 1994. Goldstein, S. March. "Quantum Theory without Observers— Part One." *PhysicsToday.* College Park, MD: American Institute of Physics, 1998.

Gollancz, Victor. *Man and God.* Boston: Houghton Mifflin Company, 1951.

Good News for Modern Man. New York: American Bible Society, 1970.

Gopal, Sarvelli. *Radhakrishnan, A Biography.* London: Unwin Hyman Ltd., 1989.

Gospel According to Thomas, The. Coptic text established and translated by A. Guillaumont, et al. San Francisco: Harper & Row, 1959.

Gospel of Thomas, The. York, England: The Ebor Press, 1987.

Gospel of Thomas Comes of Age, The. Harrisburg, PA: Trinity Press International, 1998.

Goswami, Amit. *The Self-Aware Universe.* New York: Tarcher, 1993. *Chuang-tzu: The Inner Chapters.* London: George Allen & Unwin, 1981.

Great Treasures of Ancient Teachings. 627 vols. Berkeley, CA: DharmaPublishing, 198–93.

Greenberger, Dadiel M., and Zeilinger, eds. Vol. 755, *Fundamental Problems in Quantum Theory: A Conference Held in Honor of Professor John A. Wheeler.* New York: The New York Academy of Sciences, 1995.

Greene, Brian. The Fabric of the Cosmos: Space, Time, and the Texture of Reality. New York: Vintage Books, 2004.

Gregory of Nyssa, *The Life of Moses.* Translated by Abraham J. Malherbe and Everett Ferguson. New York: Paulist Press, 1978.

Griffiths, Bede. *Return to the Center.* London: Collins Fontana, 1978.

————. *River of Compassion.* Warwick, NY: Amity House, 1987.

Guthrie, Kenneth Sylvan, comp. and trans. *The Pythagorean Sourcebook andLibrary.* Grand Rapids, MI: Phanes Press, 1987.

Hafiz. *The Gift.* Translated by Daniel Ladinsky. New York: Arkana, Penguin Group, 1999.

Haich, Elizabeth. *Sexual Energy and Yoga.* London: George Allen & Unwin Ltd. Hakeda, Yoshito, trans. *Kukai: Major Works.* New York: Columbia University Press, 1972.

Hammarskjöld, Dag. *Markings.* Translated from Swedish by Leif Sjöberg and W. H. Auden. New York: Ballantine Books, 1993.

Hardy, G. H. *Orders of Infinity, the 'Infinitärcalcul' of PaulDuBois Reymond.* Cambridge, England: Cambridge University Press, 1910.

Harman, W. Willis. "Business as a Component of the Global Economy." *Noetic Sciences Review,* November 1971.

Harman, Willis, and Hormann, John. *Creative Work: The Constructive Role of Business in a Transforming Society*. Indianapolis, IN: Knowledge Systems, 1990.

Harman, Willis and Rheingold, Howard. *Higher Creativity: Liberating the Unconscious for Breakthrough Insights*. Los Angeles: Jeremy P. Tarcher Inc., 1984.

Harold, Preston, and Babcock, Winifred. *Cosmic Humanism and World Unity*. New York: Dodd, Mead, 1971.

Haughey, John C. *The Conspiracy of God*. Garden City, NY: Image Books, 1976.

Hawking, Stephen. Singularities in the Universe, *Physical Review Letters* 17.

———. *A Brief History of Time*. New York: Bantam Books, 1988.

Hawking, S. W. and Ellis, G. F. R. *The Large Scale Structure of Space-Time*. Cambridge, England: Cambridge University Press, 1973.

Hawkins, Donald J., ed. *Famous Statements, Speeches and Stories of Abraham Lincoln*. Scarsdale, NY: Heathcote Publications, 1981.

Heath, Sir Thomas L., trans., "The Thirteen Books of Euclid's Elements." *The Great Books of the Western World*. Vol. 11. Chicago: Encyclopaedia Britannica Inc., 1952.

Heidegger, M. *The Basic Problems of Phenomenology*. Translated by Albert Hofstadter. Indianapolis: Indiana University Press, 1998.

Heisenberg, Werner. *The Physical Principles of the Quantum Theory*. New York: Dover Publications Inc., 1930.

———. *The Physicist's Conception of Nature*. New York: Harcourt and Brace, 1955.

———. Physics and Philosophy: The Revolution in Modern Science. New York: Harper & Row, 1958.

———. *Across the Frontiers*. New York: Harper & Row, 1974.

Herbert, Nick. *Quantum Reality: Beyond the New Physics*. New York: Doubleday, 1985.

Heschel, Abraham Joshua Heschel. *The Prophets*. New York: HarperCollins, 1969.

———. *I Asked for Wonder: A Spiritual Anthology*. Edited by Samuel H. Dresner. New York: Crossroad, 2000.

Hiley, Basil. "Vacuum or Holomovement." *The Philosophy of Vacuum*. Oxford: Oxford University Press, 1991.

Hirtenstein, S., ed. Journal of the Muhyiddin Ibn 'Arabi Association. Oxford: 1981.

Hodson, Geoffrey. *The Hidden Wisdom in the Holy Bible.* (An examination of the idea that the contents of the Bible are partly allegorical.) 3 vols. Wheaton, IL: Theosophical Publishing House, 1955.

———. Lecture Notes from *The School of Wisdom.* Vol. 2. Adyar, Madras, India: Theosophical Publishing House, 1955.

———.*The Brotherhood of Angels and Men.* Wheaton, IL: Theosophical Publishing House, 1983.

Horvath, T. "A Study of Man's Horizon-Creation: A Perspective for Cultural Anthropology." *The Concept and Dynamic of Culture.* Edited by B. Bernardi. The Hague: Mouton Publishers, 1976.

———. "Methods and Systematic Reflections: The Structure of Scientific Discovery and Man's Ultimate Reality and Meaning." *Ultimate Reality and Meaning* 3 (1980):–-161.

———. "John Neumann's Idea of Ultimate Reality and Meaning." *Ultimate Reality and Meaning* 20 (1997):134–7.

Huffines, LaUna. Bridge of Light: Tools of Light for Spiritual Transformation. New York: H. J. Kramer Inc., Pub., 1993.

———. Healing Yourself with Light: How to Connect with the Angelic Healers. New York: H. J. Kramer Inc., Pub., 1995.

Hughes, David. *The Star of Bethlehem Mystery. An Astronomer's Confirmation.* New York: Walter & Co., and London: Dent & Sons, 1979.

Hume, Robert Ernest, trans. *The Thirteen Principal Upanishads,* from Sanskrit, with an outline of the philosophy of the Upanishads. London: Oxford University Press, 1971.

Humes, James C. *The Wit and Wisdom of Abraham Lincoln.* New York: Gramercy Books, 1999.

Huxley, Aldous. *The Perennial Philosophy.* New York: Meridian Books, 1968.

Huxley, Julian, *Aldous Huxley 1894–1963; A Memorial Tribute.* London:Chato and Windus; New York: Harper & Row.

Hyujong. Choson sidae pyon. HangukPulgyo chonso [Comprehensive Collection of Korean Buddhism]. Vol. 7. Seoul: Tongguk Taehakkyo Ch'ulp'anbu, 1990, in classical Chinese.

————. *The Seven Days of the Heart.* Translated by Pablo Beneito and Stephen Hirtenstein. Oxford: Anqua Publishing, 2000.

Ibn 'Arabi. "Bezels of Wisdom." *Classics of Western Spirituality,* trans. by Ralph Austin. New York: Paulist Press, 1980.

Inada, Kenneth. Nagarjuna: A Translation of his Mula-madhyamaka-karika with an Introductory Essay. Tokyo: Hokuseido Press, 1970.

Irenaeus. *Adversus Haereses. The Ante-Nicene Fathers.* Vol. 1, bk. 2., ch. 28, 4. Edited by A. Roberts and J. Donaldson. Grand Rapids, MI: Eerdmans, 1958.

Jaeger, W. Two Rediscovered Works of Ancient Christian Literature: Gregory of Nyssa and Macarius. Leiden, 1954.

Jahn, R. G., and Dunne, B. J. *Margins of Reality: Role of Consciousness in the Physical World.* San Diego: Harcourt, Brace, Jovanovish, 1988.

Jaini, Padmanabh S. *The Jaina Path of Purification.* Berkeley, CA: University of California Press, 1979.

Jammer, Max. The Conceptual Development of Quantum Mechanics. New York: McGraw-Hill, 1966.

————. The Philosophy of Quantum Mechanics. New York: John Wiley, 1974.

————. Concepts of Space: The History of Theories of Space in Physics. New York: Dover Publications, 1993.

James, Joseph. *The Way to Mysticism.* London: Jonathan Cape, 1950.

James, William. *The Varieties of Religious Experience.* London: LongmansGreen, 1919.

————. *William James: The Essential Writings.* Edited by Bruce W. Wilshire. Albany: State University of New York, 1984.

————. *A Pluralistic Universe.* Lincoln: University of Nebraska, 1996.

————. *William James: The Essential Writings.* Edited by Bruce W. Wilshire. Albany: State University of New York, 1984.

Jami, Maulana Abdurrahman. *Lawa'ih.* Tehran, 1342 sh./1963.

————. *Diwan-Ikamil,* ed. Hashim Riza. Tehran, 1962.

Jeans, Sir James. *The Mysterious Universe.* Cambridge: Cambridge UniversityPress, 1931.

John of the Cross. *Dark Night of the Soul.* New York: Doubleday, 1959.

————. *The Collected Works of St. John of 'the Cross.* Translated by Kieran Kavanaughand Otilio Rodriguez. Washington DC: Institute of Carmelite Studies, 1973.

————. *Flame of Love, Spiritual Canticle.* Classics of Western Spirituality. New York: Paulist Press, 1984.

John Paul II. Encyclical Letter of John Paul II to The Catholic Bishops of the World on the Relationship Between Faith and Reason. Rome: L'Osservatore Romano, 1998.

Kalupahana, David J. *Nagarjuna: The Philosophy of the Middle Way.* Albany, NY: SUNY Press, 1986.

Kanamori, Akihiro. *The Higher Infinite.* Berlin: Springer-Verlag, 1997.

Kawai, Hayao. *The Buddhist Priest Myoe: A Life of Dreams.* Translated byMark Unno. Venice, CA: Lapis Press, 1992.

Keating, Thomas. *Open Mind, Open Heart.* Warwick, NY: Amity House, 1986.

Keel, Hee Sung. *Chinul: The Founder of the Korean Son [Zen] Tradition.* PhD diss., Harvard University, 1977.

Kempis, Thomas à. *Imitation of Christ.* New York: Doubleday, 1955.

Kerner, Fred, ed. *A Treasury of Lincoln Quotations.* New York: Doubleday &Co., 1965.

Kerrigan, A. St. Cyril of Alexandria: Interpreter of the Old Testament. Rome, 1952.

Keynes, Geoffrey, ed. *The Writings of William Blake.* 3 vols. London, 1925.

Kieffer, Gene, ed. Kundalini for the New Age: Selected Writings by Gopi Krishna. New York: Bantam Books, 1988.

King, Martin Luther Jr. *Stride Toward Freedom.* New York: Harper & Row, 1958.

Kitaro, Nishida. *Intelligibility and the Philosophy of Nothingness.* Translated by R. Schinzinger. Tokyo: Muruzen, 1958; also reprint, Westport, CT.: Greenwood Press, 1973.

Kiyota, Minoru. *Shingon Buddhism: Theory and Practice.* Los Angeles: Buddhist Books International, 1978.

————. *Last Writings: Nothingness and the Religious Worldview.* Translated by D. A. Dilworth. Honolulu: University of Hawaii Press, 1987.

Klotz, Neil Douglas. *The Hidden Gospel.* Wheaton, IL: Quest Books, 1999.

Kochumuttom, Thomas. A Buddhist Doctrine of Experience: A New Translation and Interpretation of the Works of Vasubandhu the Yogacarin. Delhi: Motilal Banarsidass, 1982.

Kodlubovsky, E. and Palmer, G.E.H. *Early Fathers from the Philokalia.* London: Faber and Faber Limited, 1954.

Koestler, Arthur. *The Act of Creation.* New York: Dell, 1964.

Krishna, Gopi. Kundalini: The Evolutionary Energy in Man. Berkeley, CA, 1970.

———. The Biological Basis of Religion and Genius. New York: Harper & Row, 1972.

———. Higher Consciousness: The Evolutionary Thrust of Kundalini. New York: The Julian Press, 1974.

———. *The Awakening of Kundalini.* New York: Dutton, 1975.

La Fleur, William R., ed. *Dogen Studies.* Honolulu: University of Hawaii Press, 1985.

Landry, Tom, with Greg Lewis. *Tom Landry: An Autobiography.* New York: HarperCollins, 1990.

Lao Tzu. *Tao Te Ching.* Translated by D. C. Lau. Baltimore: Penguin Books, 1963.

———. *The Tao Teh King: Sayings of Lao Tzu.* Translated with commentary by C. Spurgeon Medhurst. Wheaton, IL: Theosophical Publishing House, 1972.

———. *Lao Tsu: Tao Te Ching.* Translated by Gia-Fu Feng and Jane English. New York: Vintage, 1972.

———. Lao-Tzu: Te-Tao Ching: A New Translation Based on the Recently Discovered Ma-wang-tui Texts. Translated by Robert G. Henricks. New York: Ballantine, 1989.

Lavine, Shaughan. *Understanding the Infinite.* Cambridge, MA: Harvard University Press, 1994.

Lawson, John. *The Biblical Theology of Saint Irenaeus.* London: Epworth Press, 1948.

Leadbeater, C. *The Chakras.* Wheaton, IL: Theosophical Publishing House, 1969.

———. *The Hidden Side of Things.* Wheaton, IL: Theosophical Publishing House, 1974.

———. *The Science of the Sacraments.* Adyar, India: Theosophical Publishing House, 1974.

————. *The Christian Gnosis,* Ojai, CA: St. Alban's Pres, 1983.

Lebreton, J. History of the Dogma of the Trinity from its Origins to the Council of Nicaea. London: Burns, Oates, and Washbourne, 1939.

Lee, Peter H., ed. *The Silence of Love: Twentieth Century Korean Poetry.* Honolulu: The University Press of Hawaii, 1980.

————, ed. *Sourcebook of Korean Civilization.* New York: Columbia University Press, 1993.

Legge, James, trans. and ed. *The Life and Work of Mencius.* Oxford: Clarendon Press, 1895.

Lewis, Gilbert N. *The Anatomy of Science.* New Haven: Yale University Press, 1926.

Liu, Ming-wood. *The Teaching of Fa-Tsang: An Examination of Buddhist Metaphysics.* Ann Arbor, MI: University Microfilms International, 1979.

Logan, Alastair H. B. *Gnostic Truth and Christian Heresy: A Study in the History of Gnosticism.* Edinburgh, Scotland: Hendrickson Publishers, 1996.

Lombardi, Vince, with Heinz, W. C. *Run to Daylight.* Englewood Cliffs, NJ: Prentice Hall, 1963.

Lopez, Donald S., and Steven C. Rockefeller, eds. *Christ and the Bodhisattva.* Albany, NY: State University of New York Press, 1987.

Lorimer, David, ed. The Spirit of Science: From Experiment to Experience. New York: Continuum, 1999.

Lovejoy, Arthur. *The Great Chain of Being.* Cambridge, MA: Harvard University Press, 1953.

Luisi, P.L. *What Is Matter, What Is Life?* Columbia University Press, 2008.

MacGregor, G. H. C. *St. John's Gospel.* London: Hodder & Stoughton, 1936.

Mackay, Alan L. *A Dictionary of Scientific Quotations.* Bristol and Philadelphia: Institute of Physics Publishing, 1992.

Mahamudra. *Mahamudra: The Quintessence of Mind and Meditation.* Translated and annotated by Lobsang P. Lhalungpa. Boston: Shambhala Publications Inc., 1986.

Maimonides, Moses. *The Guide for the Perplexed.* Translated by M. Friedlander. New York: Dover Publications, 1956.

Manaka, Fujiko. *Jichin Kasho oyobi shugyokushu no kenkyu* [Master Jien and the Collection of Gleaned Jewels]. Kawasaki, Japan: Mitsuru Bunko, 1974.

Mandino, Og. *The Greatest Salesman in the World.* New York: Frederick Fell Publishers, 1973.

Margenau, Henry. *The Nature of Physical Reality.* New York: McGraw-Hill, 1950.

Margenau, Watson, and Montgomery. *Physics Principles and Applications.* New York: McGraw-Hill Book Company, 1953.

Mason, S.F. *A History of the Sciences.* London: Routledge, 1953.

Matt, Daniel C. *Zohar: The Book of Enlightenment.* Mahwah, NJ: PaulistPress, 1983.

———. *The Essential Kabbalah.* San Francisco: Harper San Francisco, 1996. Maurice, Nicoll. *The New Man.* Baltimore: Penguin Books, 1972.

Mayotte, Ricky Alan. *The Complete Jesus.* South Royalton, VT: SteerforthPress, 1997.

McGreal, Ian P., ed. *Great Thinkers of the Eastern World.* HarperCollins Publishers, 1995.

Mearns, David, ed. *Lincoln Papers.* 2 vols. New York: Doubleday & Co., 1948.

Mechtchild of Magdeburg. *Das Flieszende Licht der Gottheit von Mechtchild von Magdeburg* [The Flowing Light of the Godhead]. Berlin, 1909.

Mei, Yi-Pao. *The Ethical and Political Works of Motse.* Westport, CT: Hyperion Press, 1973.

Mersch, Emile. *The Whole Christ.* Milwaukee: Bruce & Bruce, 1938. Merton, Thomas. *New Seeds of Contemplation.* New York: New DirectionsBooks, 1961.

———. *The Asian Journal of Thomas Merton.* Edited by Naomi Burton, Brother Patrick Hart, and James Laughlin. New York: New Directions Press, 1973.

———. *The Way of Chuang Tzu.* New York: Norton Co., 1975.

———. *The Intimate Merton.* Edited by Patrick Hart and Jonathan Montaldo. New York: HarperCollins Publishers, 1999.

Milarepa, Jetsun. *One Hundred Thousand Songs of Milarepa.* Translated andannotated by Garma C. C. Chang. 2 vols. Boulder: Shambala, 1977.

Miller, William Lee. *Lincoln's Virtues: An Ethical Biography.* New York: AlfredA. Knopf, 2002.

Miller, W. R. *Nonviolence: A Christian Interpretation.* New York: Schocken, 1966.

Misner, C. W., Thorne, K. S., and Wheeler, J. A. *Gravitation.* San Francisco: Freeman, 1973.

Mitchell, Donald. Spirituality and Emptiness: The Dynamics of Spiritual Life in Buddhism and Christianity. New York,: Paulist Press, 1991.

Mitchel, Stephen. *The Enlightened Mind.* New York: Harper Perennial, 1993.

Montague, G. *Growth in Christ.* Kirkwood, MO: Maryhurst Press, 1961.

Moore, Gregory H. Zermelo's Axiom of Choice: Its Origins, Developments, andInfluence. New York: Springer-Verlag, 1982.

Morgan, Tom, ed. *A Simple Monk.* Novato, CA: New World Library, 2001.

Morris, James. *Introduction, to Wisdom of the Throne.* Translation of Mulla Sadra's *al-Hikmat al-'arshiyyah.* Princeton, NJ: Princeton University Press, 1981.

Muktananda, S. P. *Guru.* New York: Harper & Row, Publishers. 1971.

———. *Kundalini: The Secret of Life.* South Fallsburg, NY: SYDA Foundation, 1979.

———. *Play of Consciousness.* New York: Harper & Row, 1980.

Muller, Robert. *New Genesis: Shaping a Global Spirituality.* Garden City, NY: Image Books, 1984.

———. *What War Taught Me about Peace.* New York: Doubleday, 1985.

———. *A Planet of Hope.* Warwick, NY: Amity House, 1985.

Murray, Bruce C. *Navigating the Future.* New York: Harper & Row, Publishers, 1975.

Namgyal, Takpo Tashi. *Mahamudra:The Quintessence of Mind and Meditation.* Translated and annotated by Lobsang P. Lhalungpa. Boston: Shambhala Publications Inc., 1986.

Nanopoulos, D. V. "Tales of the Gut Age." *Grand Unified Theories and Related Topics*. Edited by M. Konuma and T. Maskawa. Proceedings of the 4th Kyoto Summer Institute 5–63. 1981.

Nasr, Seyyed Hossein. *Three Muslim Sages*. New York: Carvan Books, 1969.

———. Sadr al-Din Shirazi and His Transcendental Theosophy. Tehran: Iranian Academy of Philosophy, 1978.

Nazi Conspiracy and Aggression, Official Records of the International Military Tribunal at Nuremberg. 8 vols. New York: United Nations, 1946.

Needleman, Jacob. *Lost Christianity*. New York: Bantam Books, 1980.

Neumann, J. von. *The Mathematical Foundations of Quantum Mechanics*. Translated by R. T. Beyer. Princeton, NJ: Princeton University Press, 1955.

———. *Collected Works*. Edited by A. H. Taub. Oxford: Pergamon, 1961.

Neumann, J. von, and Morgenstern, Oskar. *The Theory of Games and Economic Behavior.* Princeton, NJ: Princeton University Press, 1959.

———. *Theory of Self-Reproducing Automata*. Urbana, Ill.: University of Illinois Press, 1966.

New American Bible, The. Translated from the Original Languages with Critical Use of All the Ancient Sources. Washington DC: Confraternity of Christine Doctrine, 1970.

Newton, I. "Mathematical Principles of Natural Philosophy." Vol. 34. *Great Books of the Western World*. Chicago: Encyclopaedia Britannica Inc., 1971.

Nicolaevsky, Boris, ed. *The Crimes of the Stalin Era*. New York: The New Leader, 1962.

Nicoll, Maurice. *The New Man*. Baltimore: Penguin Books, 1972.

Nikaya, Digha. *Thus Have I Heard: The Long Discourses of the Buddha*. Translated by Maurice Walshe. London: Wisdom Publications, 1987.

Nikilananda, Swami. *The Upanishads*. 4 vols. New York: Harper & Brothers, 1959.

Nishitani, Keiji. *Religion and Nothingness*. Translated by Jan van Bragt. Berkeley: University of California Press, 1982.

Novak, Philip. The World's Wisdom: Sacred Texts of the World's Religions. New York: HarperCollins Publishers, 1994.

Novikov, I. D. The River of Time. Cambridge: Cambridge University Press, 1998.

Nyingma Edition of the Tibetan Buddhist Canon. 120 vols. Berkeley, CA: Dharma Publishing, 1981.

O'Leary, Brian. Miracle in the Void. Kihei, Hawaii: Kamapua'a Press, 1996.

Ooms, Herman. Tokugawa Ideology: Early Constructs, 1570–1680. Princeton, NJ: Princeton University Press, 1985.

Origen. Origenes Werke. 8 vols. Leipzig, 1899–1925.

Osborne, E. F. The Philosophy of Clement of Alexandria. Cambridge: Cambridge University Press, 1957.

Osbourne, Arthur. Ramana Maharshi and the Path of Self-Knowledge. New York: Samuel Weiser, 1973.

Pagels, Elaine. The Gnostic Gospels. New York: Random House, 1980.

Pagels, Heinz. The Cosmic Code. London: Penguin Books, 1994.

Pak, Chong-hong. "Wonhyo's Philosophical Thought." Assimilation ofBuddhism in Korea: Religious Maturity and Innovation in the SillaDynasty. Edited by Lewis R. Lancaster and C. S. Yu. Berkeley, CA: Asian Humanities Press, 1991.

Pantanjali. How to Know God: The Yoga Aphorisms of Pantanjali. Translated by Swami Pascal, Blaise. Pensees and Other Writings. Translated by Honor Levi. Oxford: Oxford University Press, 1995.

Paul VI, Pope. Ecclesiam Suam. New York: Paulist Press, 1965.

Pauling, Linus. The Nature of the Chemical Bond. New York: Cornell University Press, 1960.

Peebles, P. J. E. Principles of Physical Cosmology. Princeton, NJ: Princeton University Press, 1993.

Peierls, R. E. The Laws of Nature. George Allen, 1955.

Pelikan, Jaroslav. The Idea of the University: A Reexamination. New Haven, CT: Yale University Press, 1992.

Penrose, Roger. "Gravitational Collapse and Space-Time Singularities." Physical Review Letters. 1965.

———. Cosmology. London: BBC Publications, 1974.

———. "Singularities in Cosmology," *Confrontation of Cosmological Theories with Observational Data.* Edited by Longair, M. S. Boston: D. Reidel.

———. "Singularities and Time-asymmetry." *General Relativity: An Einstein Centenary Survey.* Edited by Stephen Hawking and W. Israel. Cambridge: Cambridge University Press, 1979.

———. *Shadows of the Mind.* Oxford: Oxford University Press, 1994.

———. *The Large, the Small and the Human Mind.* Cambridge: CambridgeUniversity Press, 1997.

Petley, B. The Fundamental Physical Constants and the Frontier of Measurement, Bristol, England: Adam Hilger Ltd., 1985.

Phillips, Donald T. *Run to Win: Vince Lombardi on Coaching and Leadership.* New York: St. Martin's Press, 2001.

Planck, Max. *Where Is Science Going?* New York: Norton, 1932.

Plato. "Dialogues of Plato: Parmenides." *Great Books of the Western World.*Chicago: Encyclopaedia Britannica Inc., 1971.

———. "Timaeus." Vol. 7, *Great Books of the Western World.* Chicago: Encyclopaedia Britannica Inc., 1952.

Plotinus. "The Six Enneads." Vol. 17, *Great Books of the Western World.* Chicago: Encyclopaedia Britannica Inc., 1952.

———. *Works.* Translated by A. H. Armstrong. London: Heinemann, 1996–1984.

Poincare, H. *La Valeur de la Science.* Paris: Flammarion. 1904.

———. Revue de métaphysique et de morale 20 (1912): 486.

Popper, Karl Raimund. *The Logic of Scientific Discovery.* London: Hutchison, 1959.

———. Conjectures and Refutations: The Growth of Scientific Knowledge. NewYork: Basic Books, 1963.

———. Objective Knowledge: An Evolutionary Approach. Oxford: Clarendon Press, 1972.

Prabhupada, A. C., and Bhaktivedanta Swami. *Bhagavad-Gita as It Is.* LosAngeles: Macmillan Publishing, 1973.

———. *Sri Isopanisad.* Los Angeles: The Bhaktivedanta Book Trust, 1995. Preston, Harold. *Cosmic Humanism and World Unity.* New York: Dodd,Mead, 1971.

Prigogine, I. *From Being to Becoming.* San Francisco: Freeman, 1980.

Quine, W. V. O. *Set Theory and Its Logic.* Cambridge, MA: Harvard UniversityPress, 1963.

Qushayri, al-. *Principles of Sufism.* Translated by B. R. Von Schlegel. Berkeley,CA: Mizan Press, 1990.

Radhakrishnan, Sarvepalli. *The Bhagavadgita.* London: Allen and Unwin, 1948.

———. *The Principal Upsanishads.* New York: Harper & Brothers, 1953.

Radhakrishnan, Sarvepalli, and Moore, Charles. *A Sourcebook in IndianPhilosophy.* Princeton, NJ: Princeton University Press, 1957.

Radin, D. I. *The Conscious Universe.* San Francisco: Harper Edge, 1997.

Rahula, Walpola. *What the Buddha Taught.* New York: Grove Press, 1974.

Rama, Swami. *The Book of Wisdom.* Kanpur, India: Himalayan InternationalInstitute of Yoga, Science & Philosophy, 1972.

Ramakrishna. *Ramakrishna: Prophet of New India.* Translated by SwamiNikhilananda. New York: Harper & Brothers, n.d.

Randall, J. C. *Mr. Lincoln* 4 vols. New York: Dodd Mead, 1945.

Rayfield, Donald. *Stalin and His Hangmen.* New York: Random House, 2004.

Reichenbach, Hans. *Experience and Prediction.* Chicago: University of ChicagoPress, 1938.

———. *Philosophic Foundations of Quantum Mechanics.* Berkeley: University ofCalifornia Press, 1946.

———. *The Direction of Time.* Berkeley: University of California Press, 1956.

Reischauer, A. K. "Genshin's Ojo Yoshu: Collected Essays on Birth intoParadise." *Transactions of the Asiatic Society of Japan* 7, 2nd ser. (1930):16–97.

Reiser, Oliver. *Cosmic Humanism.* London: Schenkman, 1966.

Rele, Vasant. *The Mysterious Kundalini.* India: D. B. Taraporevala Sons & Company. Renan, Ernest. *The Life of Jesus.* Translated by C. E. Wilbur. New York: Everyman's Library, E. P. Dutton, 1987.

Ribera, Francisco de, *Vida de S. Teresa de Jesus.* Barcelona: Nuova ed., 1908.

Richards, I. A. *Mencius on the Mind.* London: Kegan Paul, Trench, Trubner & Company, 1932.

Richardson, Robert D. Jr. *Emerson*. Berkeley: University of California Press, 1995.

Rinpoche, K. *Gently Whispered: Oral Teachings*. Compiled, edited, and annotated by E. Selandia. Tarrytown, NY: Station Hill Press Inc.

Roberts, Alexander, DD, and James Donaldson, LLD, eds. "The writings of the Fathers down to CE 325." *Ante-Nicene Fathers*. Vols. 1–10. Peabody, Mass: Hendrickson Publishers, 1999.

Robinson, James M., gen. ed., *The Nag Hammadi Library in English*. San Francisco: Harper & Row, 1988.

———. "The Gospel according to Thomas." *The Nag Hammadi Library in English*. San Francisco: Harper & Row, 1988.

Roche de Coppens, Peter. *Spiritual Man in the Modern World*. Washington: University Press of America, 1976.

———. Spiritual Perspective II: The Spiritual Dimension and Implications of Love, 1984.

———. The Nature and Use of Ritual for Spiritual Attainment. St. Paul, MN: Llewellyn Publications, 1985.

———. *Apocalypse Now*. St. Paul, MN: Llewellyn, 1988.

———. *The Art of Joyful Living*. Rockport, MA: Element, 1992.

———. Divine Light and Fire: Experiencing Esoteric Christianity. Rockport, MA: Element, 1992.

———. *The Spiritual Family in the 21ˢᵗ Century*. Philadelphia: Xlibris Book Publishers, 2005.

———.Religion, Spirituality and Healthcare. Philadelphia: Xlibris, 2007.

Rodd, Laurel Rasplica. *Nichiren: Selected Writings*. Honolulu: University Pressof Hawaii, 1980.

Rotman, Brian. *Signifying Nothing*. Stanford, CA: Stanford University Press, 1987.

Rucker, Rudy. *Infinity and the Mind*. New York: Bantam Books, 1983.

Rumi, Jalal al-Din. *The Discourses of Rumi*, Translated by A. J. Arberry. London: Murray, 1961.

———. *Teachings of Rumi theMasnavi*, translated and abridged E. H. Whinfield. New York: E. P. Dutton & Co., 1975.

———. *Light upon Light*. Translated by Andrew Harvey. Berkeley: North Atlantic Books, 1996.

————. *The Essential Rumi.* Translated by Coleman Barks with John Moyne, A. J. Arberry, and Reynold Nicholson. New York: Quality Paperback Book Club, 1998.

Rump, Ariane. *Commentary on the Lao Tzu by Wang Pi.* Honolulu: University of Hawaii Press, 1979.

Russell, Bertrand. *Introduction to Mathematical Philosophy.* New York: The MacMillan Company and George Allen & Unwin Ltd., 1919.

————. *Mysticism and Logic.* Totowa, NJ: Barnes & Noble Books, 1981.

————. *Human Knowledge, Its Scope and Limits.* New York: Simon and Schuster, 1948.

————. *The Analysis of Mind.* London: George Allen and Unwin, Ltd., 1956.

————. *Wisdom of the West.* New York: Doubleday and Company Inc., 1959

Russell, Bertrand, and Whitehead, Alfred North. *Principia Mathematica.* 2nded. 3 vols. London: Cambridge University Press, 1935.

Russell, Peter. The Global Brain: Speculations on the Evolutionary Leap to Planetary Consciousness. Los Angeles: J. P. Tarcher Inc., 1983.

Sahn, S. *The Compass of Zen.* Boston: Shambhala Publications Inc., 1997.

Sandburg, Carl. *Abraham. Lincoln.* 6 vols. New York: Charles Scribner & Sons, 1943.

————. *Abraham Lincoln.* Norwalk, CT: The Easton Press, 1954.

Sanella, Lee, MD. *Kundalini—Psychosis or Transcendence?* San Francisco: Sannela, 1976.

Sanford, John A. *The Kingdom Within.* New York: Harper & Row, 1987.

Santayana, George. *Reason in Common Sense.* New York: Charles Scribnerand Sons, 1927.

Sartre, Jean-Paul. *Being and Nothingness.* Translated by Hazel Barnes. New York: Philosophical Library, 1956.

Sastri, S. S., and Raja, C. K., trans. and eds. *The Bhamati of Vacaspati.* Madras, India: Theosophical Publishing House, 1933.

Satprakashanada, Swami. *Methods of Knowledge.* Calcutta: Advaita Ashrama, 1974.

Saunders, S. and Brown, H. R, eds. *The Philosophy of Vacuum*. Oxford: Oxford University Press, 1991. Philip, ed. "A Select Library of theChristian Church." Vols. 1–28. *NiceneandPost-NiceneFathers*. Peabody, MA: Hendrickson Publishers, 1999.

———. *History of the Christian Church*. 8 vols. Peabody, MA: Hendrickson Publishers Inc., 2002.

Schilpp, Paul A, ed. *The Philosophy of Sarvepalli Radhakrishnan*. New York: Tudor Publishing Company, 1952.

Schimmel, Annemarie. *Mystical Dimensions of Islam*. Chapel Hill: The University of North Carolina Press, 1975.

Schrödinger, Erwin. *What Is Life?* Cambridge: The MacMillan Company, 1946.

Schuhmacher, Stephan. The Encyclopedia of Eastern Philosophy and Religion. Boston: Shambhala, 1994.

Schure, Edouard. *The Great Initiates*. Translated by Gloria Rasberry. San Francisco: Harper & Row, 1980.

Seife, Charles. *Zero: The Biography of a Dangerous Idea*. New York: The Penguin Group, 2000.

Sells, Michael A. "Foundations of Islamic Mysticism." *Classics of Western Spirituality*. New York: Paulist Press, 1994.

Sharma, B. N. K. *The Brahmasutras and their Principal Commentaries: A Critical Exposition*. 3 vols. Bombay: Bharatiya Vidya Bhavan. Vol. 1, 1971; vol. 2, 1974; vol. 3. 1978.

Sheldrake, R. A New Science of Life: The Hypothesis of Formative Causation. London: Blond & Briggs, 1981.

Shipov, Gennady. *A Theory of Physical Vacuum.: A New Paradigm*. Moscow: Russian Academy of Natural Sciences, 1998.

Simonetti, Manlio. Biblical Interpretation in the Early Church: An Historical Introduction to the Patristic Exegesis. Translated by John A. Hughes. Edinburgh: T&T Clark, 1994.

Singh, Gopal Singh, trans. and ed. *Sri Guru Granth Sahib: An Anthology*. Calcutta: M. P. Birla Foundations, 1989.

Singh, Maharay Charan. *Light on Saint John*. Punjab, India: Radha Soami Satsang Beas, 1985.

Skolimowski, Henry. "Global Philosophy as the Canvas for Human Unity." *The American Theosophist*. May 1983.

Skovorodá, Hryhorij Savych. *Hryhory Skovorodá: Works in Two Volumes.* Translated by M. Kashuba and W. Shewchuk. Cambridge-Kyiv: Ukr. Research Institute of Harvard University, Shevchenko Inst. of Literature, Nat. Academy of Sciences of Ukraine, 1994.

Smith, Margaret. *Rabi'a the Mystic.* Cambridge: Cambridge University Press, 1984.

Snyder, Louis L. *The War, A Concise History, 1939–45.* New York: Julian Messner Inc., 1960.

Solzhenitsyn, Aleksandr. *The Gulag Archipelago Two.* New York: Harper & Row, 1975.

Sorokin, Pitirim. *The Ways and Power of Love.* Boston: Beacon Press, 1950.

Srinivasa, K. R. *Sri Aurobindo: A Biography and a History.* 2 vols. Pondicherry, India, 1980.

Stein, Edith. *Finite and Eternal Being: An Attempt at an Ascent to the Meaning of Being.* Translated by K. F. Reinhardt. Washington DC: ICS Publications, 2002.

———. *Edith Stein Gesamtausgabe* (ESGA, The Complete Edition of Works of Edith Stein). Edited by M. Linssen, O. C. D., and H. B. Gerl-Falkovitz. Wien: Herder, 2000.

Stulman, Julius. *Evolving Mankind's Future.* Philadelphia: J. B. Lippincott, 1967.

Sullivan, Lawrence E. Icanchu's Drum: An Orientation to Meaning in SouthAmerican. Religions. New York: Macmillan Publishing Company, 1988.

Suzuki, D. T. *An Introduction to Zen Buddhism.* New York: Grove Press, 1964.

———. *The Field of Zen.* New York: Harper & Row 1970.

———. *Mysticism: Christian and Buddhist.* Westport, CT: Greenwood, 1976.

Svarney, Patricia Barnes. Editorial Director, *Science Desk Reference.* New York:Macmillan, 1995.

Swan, Laura. *The Forgotten Desert Mothers.* New York: Paulist Press, 2001.

Symeon the New Theologian. *Hymns of Divine Love.* Translated by GeorgeA. Maloney, SJ. Denville, NJ: Dimension Books, 1968.

Tagore, Rabindranath. *The Religion of Man*. London: Allen & Unwin, 1931.

———. *Gitanjali*. New York: Macmillan Publishing Co. Inc., 1973.

Taimni, I. K. *Man, God and the Universe*. Wheaton, IL: The TheosophicalPublishing House, 1969.

Talbot, George Robert. *Electronic Thermodynamics*. Los Angeles: Pacific State University Press, 1973.

———. *Philosophy and Unified Science*. 2 vols. Madras, India: Ganesh & Company, 1977.

Tarada, Toru, and Yaoko, Mizuno, eds. *Dogen*. 2 vols. Tokyo: Iwanami, 1971.

Tarski, Alfred. *Introduction to Logic*. London: Oxford University Press Inc., 1941.

Tatia, Nathmal. *Studies in Jaina Philosophy*. Banares, India: Jaina Cultural Research Society, 1951.

Tauler, Johan. *Theologica Germanica*. Translated by Winkworth. London, 1937.

Taylor, Edwin F., and Wheeler, John A. *Space Time Physics*. San Francisco: W.H. Freeman and Co., 1966.

Taylor, L. H. *The New Creation*. New York: Pageant Press, 1958.

Tehrani, Kazem. *Mystical Symbolism in Four Treatises of Suhrawardi*. PhDdiss., Columbia University, 1974.

Teller, P. "Relativity, Wholeness, and Quantum Mechanics." *British Journal for the Philosophy of Science* 37 (1986): 71–81.

Templeton, John Marks. *The Humble Approach: Scientists Discover God*. New York: Continuum, 1995.

Teres, Gustav, SJ. *The Bible and Astronomy: The Magi and the Star in the Gospel*. Budapest: Springer Orvosi Kiado Kft., 2000.

Teresa, Mother. *A Simple Path*. Compiled by Lucinda Vardey. New York: Ballantine Books, 1995.

———. *The Joy in Loving*. New York: Viking Penguin, 1997.

Teresa of Avila. *The Interior Castle*. Translated and edited by E. Allison Peers. Garden City, NY: Image Books, 1944.

Thackston, W. M. *Mystical and Visionary Treatise of Suhrawardi*. London: The Octagon Press, 1982.

Theophilus of Antioch. "Thoeophilus to Autolycus." Translated by Marcus Dods, A. M. Vol. 2, *Fathers of the Second Century*. Peabody, Mass: Hendrickson Publishers.

Thibaut, George, trans. "The *Vedanta Sutras* with the Commentary of Ramanuja." In *The Sacred Books of the East*. Vol. 48. Oxford: The Clarendon Press, 1904.

Thomas, D. J. "The Gospel of Thomas." *The Nag Hammandi Library in English*. San Francisco: Harper & Row, Publishers, 1988.

Thomson, E. J. *Rabindranath Tagore: Poet and Dramatist*. London: Oxford University Press, 1948.

Tiller, William A. Science and Human Transformation: Subtle Energies, Intentionality and Consciousness. Walnut Creek, CA: Pavior Publishing, 1997.

Tiller, William A., Dibble, Walter E. Jr., and Kohane, Michael J. *Conscious Acts of Creation: The Emergence of a New Physics*. Walnut Creek, CA: Pavior Publishing, 2001.

Tirtha, Swami Mahjaraj Vishnu. *Devatma Shakti,* India. Tishby, Isaiah, and Lachower, Fischel. *The Wisdom of the Zohar*. New York: Oxford University Press, 1989.

Tollinton, R. B. *Clement of Alexandria*. Vols. 1 and 2. London: Williams and Norgate, 1914.

Tolstoy, Leo. Tolstoy's Writings on Civil Disobedience and Nonviolence. New York: New American Library, 1968.

Tower, Courtney. "Mother Theresa's Work of Grace." *Reader's Digest*. December 1987.

Toynbee, Arnold, and Ikeda, Daisaku. *Choose Life: A Dialogue*. London: Oxford University Press, 1976.

Tranter, Gerald. *The Mystery Teachings and Christianity*. Wheaton, IL: Theosophical Publishing House, 1969.

Trump, Ernest. *The Adi Granth, or the Holy Scriptures of the Sikhs*. New Delhi: Munshiram Manoharlal, 1970.

Trungpa, Chögyam. *Orderly Chaos: The Mandala Principle*. Boston: Shambhala, 1991.

Tseu, Augustinus A. *The Moral Philosophy of Mo-tzu*. Taiwan: Fu Jen Catholic University Press, 1965.

Tyler, Royall, trans. *Selected Writings of Suzuki Shosan.* Cornell University East Asia Papers, no. 13. Ithaca, NY: China-Japan Program, Cornell University, 1977.

Udanavarga. *The Dhammapada with the Udanavarga,* ed. by Raghavan Iyer. The Pythagorean Sangha, 1986.

Ueda, Yoshifumi, and Hirota, Dennis. *Shinran: An Introduction to His Thought.* Kyoto: Hongwanji International Center, 1989.

Underhill, Evelyn. Mysticism: A Study in the Nature and Development of Man's Spiritual Consciousness. New York: E. P. Dutton & Co. Inc., 1961.

Upanishads, The. *The Thirteen Principal Upanishads.Translated* by Robert Ernest Hume. London: Oxford University Press, 1971.

Utke, A. "The Cosmic Holism Concept: An Interdisciplinary Tool in the Quest for Ultimate Reality and Meaning." *Ultimate Reality and Meaning,* 9 (1986):134–55.

U.S. Congress House Committee on Un-American Activities. *The Crimes of Khrushchev: Hearings.* Washington DC: Government Printing Office, 1959.

U.S. Congress Senate Committee on Foreign Relations. *The Genocide Convention: Hearings before a Subcommittee,* January 23-February 9, 1950, on Executive Order. Washington DC: Government Printing Office, 1950.

U.S. Congress Senate Committee on Judiciary. *Soviet Empire: Prison House of Nations and Races.* Washington DC: Government Printing Office, 1958.

———. Soviet Empire: A Study in Genocide, Discrimination and Abuse of Power. Washington DC: Government Printing Office, 1958.

Vatican II, The Documents of, with notes by Protestant and Orthodox authorities. Edited by Walter M. Abbot, and Geoffrey Chapman, 1966.

Verster, F. "Silence, Subjective Absence and the Idea of Ultimate Reality and Meaning in Beethoven's Last Piano Sonata, Op. 111." *Ultimate Reality and Meaning* 22 (1999): 4–23.

Vivekananda. *Living at the Source.* Boston: Shambhala, 1993.

da Vinci, Leonardo. *The Notebook.* Translated and edited by E. Macurdy. London, 1954.

Vishnu Tirtha, Swami Maharaj. *Devarma Shakti.* India.

Waldenfels, Hans. *Absolute Nothingness: Foundations for a Buddhist-Christian Dialogue.* Translated by James W. Heisig. New York: Paulist Press, 1980.

Waley, Arthur. The Way and Its Power: A Study of the Tao Te Ching and Its Place in Chinese Thought. New York: Evergreen, 1958.

Wallace, A. The Taboo of Subjectivity: Towards a New Science of Consciousness. Oxford University Press, 2000.

———. *Buddhism and Science: Breaking Down New Ground.* New York: Columbia University Press, 2003.

Warder, A. K. *Indian Buddhism.* Delhi: Motilal Banarsidass, 1980.

Watson, Burton. *Chuang Tzu: Basic Writings.* New York: Columbia UniversityPress, 1964.

———. Basic Writings of Mo Tzu, Hsun Tzu, and Han Fei Tzu. New York: Columbia University Press, 1967.

———. *Complete Writings of Chuang Tzu.* New York: Columbia University Press, 1968.

Watts, Alan. The Book: On the Taboo against Knowing Who You Are. New York: Random House, 1972.

Weber, Max. *Protestant Ethic and the Spirit of Capitalism.* New York: Charles Scribner's Sons, 1976.

Weinberg, Steven. Gravitation and Cosmology: Principles and Applications of the General Theory of Relativity. New York: John Wiley, 1972.

———. The First Three Minutes: A Modern View of the Origin of the Universe. New York: Basic Books, 1976.

———. *Dreams of a Final Theory.* New York: Pantheon Books, 1992.

Weingart, R. "Making Everything Out ofNothing." *The Philosophy ofVacuum.*Oxford: Oxford University Press, 1991.

Weizsacker, C. F. von. *The Unity of Nature.* Translated by Francis J. Zucker. New York: Farrar, Strauss, Giroux, 1980.

———. *The Unity of Physics in Quantum Theory and Beyond.* Edited by Ted Bastin. Cambridge: Cambridge University Press, 1971.

Weyl, Hermann. *Philosophy of Mathematics and Natural Science.* Princeton, NJ: Princeton University Press, 1949.

———. *Theory of Groups and Quantum Mechanics.* Translated by H. P. Robertson. New York: Dover, 1950.

————. *Space Time Matter.* Translated by H. L. Brose. New York: Methuen, 1922; repr. Mineola, NY: Dover Publications Inc., 1950.

————. The Continuum: A Critical Examination of the Foundation of Analysis. New York: Dover Publications Inc., 1987.

Wheeler, John Archibald. *Frontiers of Time.* Amsterdam: North Holland, 1979.

————. *At Home in the Universe.* Woodbury, NY: American Institute of Physics, 1994.

Wheeler, J. A., and Zurek, W H., eds. *Quantum Theory and Measurement.* Princeton, NJ: Princeton University Press, 1983.

White, John, ed. *The Highest State of Consciousness.* New York: Doubleday, 1972.

————. *Kundalini, Evolution and Enlightenment.* New York: Anchor Books/ Doubleday, 1978.

————. *What Is Enlightenment?* Los Angeles: Jeremy P. Tarcher Inc.: St. Paul Minnesota, 1995.

Whitehead, Alfred North. *Process and Reality.* New York: The Macmillan Company, 1929.

————. *Adventure of Ideas.* New York: Macmillan Company, 1933.

————. *Essays in Science and Philosophy.* New York: Philosophical Library, 1947.

————. *An Introduction to Mathematics.* London: Oxford University Press, 1984.

Whitehead, Alfred North, and Bertrand Russell. *Principia Mathematica.* 3 vols. Cambridge: Cambridge University Press. Vol. 1, 1910; vol. 2, 1912; vol. 3, 1913.

Whitman, Walt. *Leaves of Grass.* New York: Quality Paperback Book Club, 1992.

Whitrow, G. J. *What Is Time?* London: Thames and Hudson, 1972.

————. *The Natural Philosophy of Time.* Oxford: Oxford University Press, 1980.*Time in History.* Oxford: Oxford University Press, 1989.

Wick, David. The Infamous Boundary: Seven Decades of Controversy in Quantum Physics.Boston: Birkhäuser, 1995.

Wigner, Eugene P. "The Limits of Science," *Proceedings of the American. Philosophical Society* 94 (1950).

Wilber, Ken. *The Atman. Project.* Wheaton, IL: Theosophical Publishing House, 1980.

————. *Quantum Questions.* London: New Science Library, 1984.

————. *Sex, Ecology and Spirituality.* Boulder, CO: Shambhala, 1995.

Wilhelm, M. *Education: The Healing Art.* Houston, TX: Paideia Press, 1995.

Wilkenhause, A. *Pauline Mysticism.* Freiburg: Herder, 1956.

Williams, R. *Jaina Yoga.: A Survey of the Mediaeval Sravkacaras.* London: Oxford University Press, 1963.

Wilson, Peter Lamborn. *Fakhruddin Iraqi: Divine Flashes,"* New York: Paulist Press, 1982.

Wingate. *Tilling the Soul.* Santa Fe, NM: Aurora Press, 1984.

Yampolsky, Philip, ed. and trans. The Platform Sutra of the Sixth Patriarch: The Text of the Tunhuang Manuscript. New York: Columbia University Press, 1967.

Yogananda, Paramahansa. *Men's Eternal Quest.* Los Angeles: Self-Realization Fellowship, 1982.

Yusuf'Ali, 'Abdullah. *The Meaning of the Holy Qur'an.* Beltsville, MD: Amana Publications, 1994.

Zaehner, R. C. Matter and Spirit: Their Convergence in Eastern Religions, Marx, and Teilhard de Chardin. New York: Harper & Row, 1963.

————. The Bhagavad Gita: with a commentary based on original sources. London: Oxford University Press, 1969.

Zajonc, A. Catching the Light: The Entwined History of Light and Mind. New York: Bantham Books, 1993.

————. *The New Physics and Cosmology.* Oxford University Press, 2006.

Zuzuki, Daisetz Teitaro. *Studies in Zen Buddhism.* London, 1927.

————. Collected Writings on Shin Buddhism. Kyoto: Shinshu Otaniha, 1973.

INDEX

PERMISSIONS ACKNOWLEDGMENTS

Grateful acknowledgment is made to the following for permission to reprint previously published material:

Excerpts from the New American Bible (NAB) circa 1970 used herein by permission of the Confraternity of Christian Doctrine (CCD), copyright owner: No permission is required for use of less than 5,000 words of the NAB in print, sound, or electronic formats provided that such use comprises less than 40% of a single book of the Bible and more than 40% of the proposed work.

Excerpts from the Good News Translation (GNT), (formerly Today's English Version) (TEV) circa 1970 used herein by permission of the American Bible Society (ABS), copyright owner: No permission is required for use of less than 500 verses in total the GNT in print, sound, or electronic formats provided that such use comprises less than 50% of a complete book of the Bible and more than 25% of the new work.

ABOUT THE AUTHOR

Orest J. Bedrij was educated in Austria and the USA. He is a multidisciplinary research scientist. His focus in life has been on the Eternal Architect and the Creator of the Universe, which he unifies here in the laws of physics. Bedrij has been researching the Cosmic Consciousness of humanity and the development of *spiritual insight* by way of *direct access* to God's essential nature. In this work, he shines light on the following:

The forbidden facts. (a) God Himself takes on the human form and lives in the world as us; (b) the Tree of Life and Knowledge from which you can eat now; (c) milk / solid food in Christianity; (d) you are God hiding from yourself in human form.

Mining the mind of God. (a) Infused contemplation; (b) beyond the Ten Commandments; (c) putting on the mind of Christ; (d) forward thinking in your living, education, and corporate life.

Your miracle after miracle life. (a) God does not need advisers; (b) "It is not I but God in me" core of miracles; (c) karma/dharma.

Enlightenment. (a) Realization of your identity; (b) we have come from the Light; (c) the Light has originated through itself; (d) the One-in-all and all in One; (e) the Supreme Being; (f) Kundalini; (g) the Holy Spirit; (h) the living water; (i) the oil of gladness.

The ultimate principle. (a) God the Father; (b) '**1**'; (c) higher and greater than the laws of nature; (d) the Great Mystery.

Cosmic Consciousness. (a) The peace that passes all understanding. (b) Fifty nations register. (c) Four hundred six individuals listing. (d)

This world is full of God, and everything you see is God. (e) The universe is the body of God.

The union with God. (a) The source of all happiness, (b) all beauty, (c) all goodness, and (d) all truth.

www.ingramcontent.com/pod-product-compliance
Lightning Source LLC
Chambersburg PA
CBHW021354210526
45463CB00001B/100